Classification and Nomenclature of Viruses

Fourth Report of the International Committee on Taxonomy of Viruses

Report prepared by the International Committee on Taxonomy of Viruses and published for the Virology Division of the International Union of Microbiological Societies (IUMS)

R.E.F. Matthews

Department of Cell Biology,
University of Auckland, New Zealand

4 figures and 3 tables, 1982

Reprint from Intervirology, Vol. 17, No. 1-3 (1982)

Previously published:

Wildy, P.:
Classification and Nomenclature of Viruses
First Report of the International Committee on Nomenclature of Viruses
Monographs in Virology, Vol. 5 (Karger, Basel 1971)
VII + 82 p.
ISBN 3-8055-1196-5

Fenner, F.:
Classification and Nomenclature of Viruses
Second Report of the International Committee on Taxonomy of Viruses
Intervirology, Vol. 7, No. 1-2 (Karger, Basel 1976)
115 p.
ISBN 3-8055-2418-8

Matthews, R.E.F.:
Classification and Nomenclature of Viruses
Third Report of the International Committee on Taxonomy of Viruses
Intervirology, Vol. 12, No. 3-5 (Karger, Basel 1979)
160 p.
ISBN 3-8055-0523-X

S. Karger · Medical and Scientific Publishers · Basel · München · Paris · London · New York · Tokyo · Sydney

Cataloging in Publication.
International Committee on Taxonomy of Viruses Classification and nomenclature of viruses.
Report prepared by the International Committee on Taxonomy of Viruses, and published for the Virology Division of the International Union of Microbiological Societies (IUMS), R.E.F. Matthews. –
Basel: Karger 1982.
Reprint from Intervirology, Vol. 17, No. 1-3, 1982.

S. Karger AG, P.O. Box, CH-4009 Basel (Switzerland)
Printed in Switzerland by
Graphische Betriebe Coop Schweiz, Basel
ISBN 3-8055-3557-0

Contents

Preface

The main section of this report summarizes the state of approved virus taxonomy and nomenclature as it was following the Plenary meeting of the International Committee on Taxonomy of Viruses (ICTV) at Strasbourg in August 1981. There are now 54 approved families and groups of viruses. In addition, I have included five proposed or possible families or groups. The data on the viruses, the lists of member viruses, and the literature references have been updated. The information given under the 'Main characteristics' section has been substantially expanded for many families and groups. For convenience, the data are presented in a standard order under a series of headings. Four pages of outline drawings are included as a visual index to remind the reader of the main morphological features of the various families and groups.

I have included a new section summarizing the procedures involved in the initiation and processing of new taxonomic proposals by ICTV for the coming 3 years. I hope that this will prove useful to new members of subcommittees and study groups.

I wish to thank all members of the Executive Committee of ICTV for their help and support, especially the Secretaries, Dr. *J. Maurin* and Dr. *V. Valenta*, and the chairmen of subcommittees, Dr. *H.-W. Ackermann*, Dr. *J. G. Atherton*, Dr. *R.I.B. Francki*, Dr. *M. Hollings*, Dr. *F.A. Murphy*, Dr. *T.W. Tinsley* and Mr. *J.F. Longworth*. ICTV could not progress without their enthusiasm and their knowledge of both viruses and virologists.

The virus descriptions came from the work of many members of Study Groups and their colleagues, too numerous to name; I thank them all most sincerely for their help and support for the work of ICTV.

For financial support for individual members to attend the meetings of the Executive Committee of ICTV, I wish to thank the following organizations: University of Missouri *(A. Eisenstark)*; Clive and Vera Ramaciotti Foundation and University of Queensland, Australia *(J. G. Atherton)*; The Royal Society, London *(K.W. Buck)*; 'Centre National de la Recherche Scientifique', Paris *(M. Bergoin)*; The Royal Society of New Zealand and Department of Scientific and Industrial Research, New Zealand *(J.F. Longworth)*; University of Adelaide, Australia *(R.I.B. Francki)*; Agricultural University, Wageningen, The Netherlands *(J. P. H. van der Want)*; Slovak Academy of Sciences, Bratislava, Czechoslovakia *(V. Valenta)*; University of Auckland Research Committee and the Royal Society of New Zealand *(R.E.F. Matthews)*; Virology Division, IUMS *(F.A. Murphy)*; Department of Virology, Karolinska Institute, Stockholm *(E. Norrby)*. In addition I thank The Royal Society of London for providing committee room facilities for the mid-term meeting of the Executive Committee in 1980.

I also wish to thank the Clive and Vera Ramaciotti Foundation, Australia, for a grant to Dr. *J. G. Atherton* of $A 23,000 to assist in the development of a computer-based virus data storage and retrieval system; and the Natural Sciences Engineering and Research Council of Canada for a grant of Can $23,000 per year for 3 years to set up a phage reference center. In due course, work carried out under these grants will be of considerable assistance in the activities of ICTV.

'*Intervirology*' is the journal of the Virology Division of the International Union of Microbiological Societies. Publication of this report as a special issue of '*Intervirology*' follows previous practice. I thank the Editor-in-Chief, Professor *J.L. Melnick*, for his help during the production process.

Auckland, 1981 *R.E.F. Matthews*

List of Officers and Members of the International Committee on Taxonomy of Viruses

Life Members	Sir Christopher Andrewes	
	F.J. Fenner	
	F.O. Holmes	
	A. Lwoff	
	R.E.F. Matthews	
	H.G. Pereira	
	P. Wildy	
	V.M. Zhdanov	

Executive Committee	**1978–1981**	**1981–1984**
President	R.E.F. Matthews	F. Brown
Vice-President	H.G. Pereira	J.P.H. van der Want
Secretaries	J. Maurin	J. Maurin
	V. Valenta	V. Valenta
Elected Members	M. Bergoin	M. Bergoin
	F. Brown	T.H. Graf
	A. Eisenstark	D.C. Kelly
	H. Fraenkel-Conrat	M.B. Korolev
	W.K. Joklik	S. Matsumoto
	J.F. Longworth	J.A. Mayo
	E. Norrby	A.F. Murant
	J.P.H. van der Want	L. van Vloten-Doting
	and the Chairmen of Subcommittees	

National Members	
Arab Republic of Egypt	M. Ouf
Argentina	M. Weissenbacher
Australia	P. Cooper
Austria	W. Frisch-Niggemeyer
Brasil	O.A. de Carvalho Pereira
Bulgaria	P. Andonov
Canada	M. Chernesky
Chile	G. Contreras
Czechoslovakia	B. Korych
Federal Republic of Germany	H.J. Eggers
Finland	E. Tapio
France	A. Kirn
Greece	A. Tsotsos
Hungary	I. Nász
India	N.P. Gupta – until 1979
	K. Banerjee – since 1979
Israel	Y. Becker
Italy	M. La Placa

National Members (continued)

Japan	
Society of Japanese Virologists	H. Uetake
Phytopathological Society of Japan	C. Matsui
Japanese Society of Veterinary Science	S. Konishi
Korea	Y.T. Yang
Mexico	C. Fernández-Tomás
Morocco	A. Chabaud
Netherlands	J. van der Veen – until 1981
	M.C. Horzinek – since 1981
New Zealand	A.R. Bellamy
Nigeria	A. Fabiyi
Norway	G. Haukenes
Peru	R. Mendes
Philippines	R. Bozeman Rodriquez
Poland	M. Morzycka
Republic of South Africa	D.W. Verwoerd
Roumania	N. Cajal
Spain	R. Nájera
Sweden	E. Norrby
Turkey	E.T. Cetin
United Kingdom	
Society of General Microbiology	H.G. Pereira – until 1980
	B.D. Harrison – since 1980
Association of Applied Biologists	M. Hollings – until 1981
	A.A. Brunt – since 1981
Uruguay	R. Somma-Moreira
USA	H.S. Ginsberg
USSR	S. Ya. Gaidamovich
Venezuela	J. Esparza
Yugoslavia	D. Sutić

Bacterial Virus Subcommittee	**1978–1981**	**1981–1984**
Chairman	H.-W. Ackermann	A. Eisenstark
Vice-Chairman	A. Eisenstark	H.-W. Ackermann
	L. Berthiaume	C.M. Calberg-Bacq
	F. Grimont	J.N. Coetzee
	L.A. Jones	F. Grimont
	A. Liss	J. Maniloff
	J. Maniloff	J.A. Mayo
	J.A. Mayo	K. Mise
	D.C. Reanney	E. Nagy
	R. Safferman	R. Schäfer
	T. Sozzi	J.-F. Vieu
	J.-F. Vieu	

Fungal Virus Subcommittee	**1978–1981**	**1981–1984**
Chairman	M. Hollings	K.W. Buck

Fungal Virus Subcommittee (continued)

	H.-W. Ackermann	H.-W. Ackermann
	R.F. Bozarth	R.F. Bozarth
	K.W. Buck	J. Bruenn
	R.M. Lister	Y. Koltin
	C.J. Rawlinson	C.J. Rawlinson
	R. Ushiyama	R. Ushiyama
	H.A. Wood	H.A. Wood

Invertebrate Virus Subcommittee	**1978–1981**	**1981–1984**
Chairman	T.W. Tinsley	J.F. Longworth
	M. Bergoin	M. Bergoin
	S.M. Gershenzon	P. Faulkner
	R.R. Granados	R.R. Granados
	C. Ignoffo	C. Ignoffo
	D.C. Kelly	D.C. Kelly
	E. Kurstak	D.L. Knudson
	J.F. Longworth	E. Kurstak
	C.C. Payne	N. Moore
	M.D. Summers	C.C. Payne
	C. Vago	M.D. Summers

Plant Virus Subcommittee	**1978–1981**	**1981–1984**
Chairman	R.I.B. Francki	R.I. Hamilton
	K.R. Bock	M. Bar Joseph
	A.A. Brunt	A.A. Brunt
	J.R. Edwardson	J.R. Edwardson
	R.I. Hamilton	R.M. Goodman
	H.T. Hsu	H.T. Hsu
	R. Hull	R. Hull
	R. Koenig	R. Koenig
	O. Lovisolo	G.P. Martelli
	G.P. Martelli	R.G. Milne
	G. Matsui	A.F. Murant
	A.F. Murant	J.W. Randles
	W.F. Rochow	M.H.V. van Regenmortel
	A. Varma	E. Shikata
	V. Veerisetty	J.H. Tremaine
		L. van Vloten-Doting

Vertebrate Virus Subcommittee	**1978–1981**	**1981–1984**
Chairman	F.A. Murphy	D.W. Kingsbury
Vice-Chairman	F. Brown	D.H.L. Bishop
	Sir Christopher Andrewes	S. Gardner
	P. Brès	A.P. Kendal
	J. Casals	J. Maurin
	J.H. Gillespie	W.E. Rawls

Vertebrate Virus Subcommittee (continued)

	I.H. Holmes	B. Roizman
	M.C. Horzinek	R.R. Rüeckert
	W.K. Joklik	F.L. Schaffer
	R. Kono	S. Siddell
	J. Maurin	H.E. Varmus
	J.L. Melnick	G. Wadell
	P.K. Vogt	E.G. Westaway
	V.M. Zhdanov	

Code and Data Subcommittee (since August 1981 called the 'Standing Subcommittee for Virus Data').	**1978–1981**	**1981–1984**
Chairman	J.G. Atherton	P. Bachmann
	H.-W. Ackermann	H.-W. Ackermann
	P. Bachmann	J.G. Atherton
	A. Eisenstark	A.J. Gibbs
	R.I.B. Francki	N. Karabatos
	F.A. Murphy	D. Knudson
	C. Vago	

Coordination Subcommittee	**1978–1981**	**1981–1984**
Chairman	R.E.F. Matthews	F. Brown
	H.-W. Ackermann	D. Baxby
	J.G. Atherton	P. Bachmann
	R.I.B. Francki	K.W. Buck
	M. Hollings	P. Dobos
	F.A. Murphy	A. Eisenstark
	T.W. Tinsley	R.I. Hamilton
		W.K. Joklik
		D.W. Kingsbury
		J.F. Longworth
		P. Scotti
		R.E. Shope
		G. Siegl
		D. Willis

President's Report

Contents

I. Introduction

The development of an internationally agreed taxonomy for viruses is detailed in the three previous reports of the International Committee on Taxonomy of Viruses (ICTV) and its predecessor, the International Committee on Nomenclature of Viruses [*Wildy*, 1971; *Fenner*, 1976; *Matthews*, 1979]. This report incorporates 51 new taxonomic proposals approved by ICTV at its meeting held during the Fifth International Congress of Virology at Strasbourg in August 1981. In addition, the data for all the approved virus families and groups have been brought up-to-date with respect to the main characteristics of the viruses, the list of member viruses, and literature references. In a report of this size it is not possible to provide extensive reference lists. References have been chosen to provide ready access to: (i) very recent papers, (ii) important review papers, and (iii) occasional papers that could be difficult to locate. The overall form of this report follows that of the Third Report.

In the Third Report I devoted some space to a discussion of the problem of 'species' in virus taxonomy. Highlights in the work of ICTV over the past 3 years relate to this question. ICTV has clarified and expanded the rules concerning the naming of virus species (p. 23) and has developed guidelines for virologists or study groups wishing to delineate and name virus species (p. 24). The first virus species, and names for them, were given provisional approval at Strasbourg. These are in the family ***Adenoviridae*** (p. 59).

II. Composition of ICTV and Its Executive Committee

A. National Members of ICTV

The rules of the International Union of Microbiological Societies (IUMS) allow each National Society that adheres to IUMS to nominate one member to ICTV. The National Membership for the period 1978–1981 is given on p. 7. As in previous years, there has been considerable difficulty in obtaining nominations from some National Societies. In order to ensure that the decisions of ICTV fairly reflect international opinion, it is most important that the officers of National Societies use their right to take part in the work of ICTV by (i) nominating a person for membership and (ii) nominating a voting proxy if that member is unable to attend the ICTV meeting which will take place in conjunction with the Sixth International Congress of Virology to be held in Sendai, Japan, in August 1984.

B. Chairmen of Subcommittees

The chairmen of subcommittees are elected by the Executive Committee of ICTV (ECICTV). The membership of these subcommittees must also be confirmed by ECICTV. The new procedure introduced during the last term has again worked effectively to minimize delays in the operation of the subcommittees. New chairmen were elected for all subcommittees at the mid-term meeting of ECICTV in London in April 1980. Membership of the new subcommittees was approved in Strasbourg in August 1981. Membership lists are given on pp. 8–10.

C. Elected Members of the Executive Committee

At the meetings in Strasbourg, August 1981, the following members retired, having served two terms: *R.E.F. Matthews* as President; *H.G. Pereira* as Vice-President; *A. Eisenstark*, *H. Fraenkel-Conrat*, *E. Norrby* and *J.P.H. van der Want*. *W.K. Joklik* retired after serving one term. *J.F. Longworth* had been elected Chairman of the Invertebrate Virus Subcommittee. *F. Brown* was elected President; and *J.P.H. van der Want* Vice-President. Thus, there were seven vacancies for elected members of the Executive Committee. *T.H. Graf*, *D.C. Kelly*, *M.B. Korolev*, *S. Matsumoto*, *J.A. Mayo* and *A.F. Murant* were elected. For the seventh vacancy on the Committee, *L. van Vloten-Doting* was elected following a postal ballot.

D. Ex officio Members

In order to ensure adequate coordination with the work of the World Health Organization (WHO) in vertebrate virology, the Executive Committee agreed that the Director of the WHO Center for the Collection and Evaluation of Data on Comparative Virology in Munich, should be ex officio a member of the Executive Committee of ICTV.

E. Life Members of ICTV

News was received that a life member, *V.L. Ryzhkov*, had died. Dr. *Ryzhkov* had been associated with the founding of ICNV in Moscow in 1966. The retiring President *R.E.F. Matthews* was elected a life member of ICTV. *H.G. Pereira* was also elected a life member in recognition of his many contributions to the development of viral taxonomy. He had been involved with the work of ICTV and its predecessor ICNV since its inception in Moscow in 1966.

F. The Present State of ICTV Membership

Since the Fourth International Congress for Virology in 1978, the membership of ICTV has increased slightly from 118 to 128. This increase has been mainly due to an enlargement of some subcommittees. In August 1981 the membership was made up as follows:

Executive Committee members	12 (plus chairmen of 6 subcommittees)
Life members	8
National members	42 (39 countries)
Subcommittee (SC) members	
Vertebrate Virus SC	14
Invertebrate Virus SC	11

Plant Virus SC	15
Bacterial Virus SC	12
Fungal Virus SC	8
Code and Data SC	6

III. Translations of the Third Report

A French translation of the Third Report of ICTV by *J. Maurin* is available from Masson SA, 120, bld. Saint-Germain, 75280 Paris Cedex 06 (France).

A Spanish translation supervised by *R. Najera* is published by the Virology Group of the Spanish Society for Microbiology and distributed in Spain by Microbiological Associates Inc.

It is hoped that similar translations of the present report will be available in due course.

IV. Work of the Subcommittees of ICTV

The host-oriented subcommittees dealing with viruses infecting bacteria, fungi, invertebrates, plants or vertebrates are the key units involved in generating new taxonomic proposals for consideration by ICTV. In addition, the Coordination Subcommittee has responsibility for viruses infecting more than one kind of host. Five families so far fall into this category – ***Poxviridae, Reoviridae, Rhabdoviridae, Parvoviridae*** and ***Iridoviridae.***

In the sections below I summarize the work of these subcommittees over the period 1978 to 1981. The 51 specific taxonomic proposals approved by ICTV are incorporated in the descriptions of the families and groups of viruses. These proposals were summarized recently in a short publication [*Matthews*, 1981].

A. Bacterial Virus Subcommittee

Eight genus names for cubic, filamentous and pleomorphic phages, and two family names for tailed phages were approved.

Members of the subcommittee have published or submitted for publication several further papers relating to bacterial virus taxonomy.

A survey completed by Dr. *H.-W. Ackermann* in January 1981 listed about 2,100 phages of known morphology not counting phage-like 'bacteriocins' and defective phages. Over 2,000 are tailed phages. Only 109 (5%) are cubic, filamentous or pleomorphic. Table I shows the numbers of phages which belong or probably belong to the ten bacterial virus families.

Most phages have been described from a relatively few bacterial groups, particularly enterobacteria (562), *Bacillus, Streptococcus* and *Pseudomonas.*

The number of phage descriptions was about 150 in 1978. The annual rate has fallen,

Table I. Numbers of phages belonging to (or probably belonging to) the ten families of bacterial viruses

Nucleic acid	Family	Number of phages
DNA	***Myoviridae***	603
	Styloviridae	1,050
	Podoviridae	372
DNA	***Microviridae***	26
	Corticoviridae	2?
	Tectiviridae	8
RNA	***Leviviridae***	34?
	Cystoviridae	1
DNA	***Inoviridae***	
	Genus *Inovirus*	20
	Genus *Plectrovirus*	13
	Plasmaviridae	3

and is now around 100. Most new descriptions are still for phages of bacteria of medical or veterinary interest. Many 'new' phages have been described for streptococci used in the fermentation industry. No phages requiring the designation of a new family have been found.

Classification of cubic, filamentous, and pleomorphic phages is relatively easy, because the groups contain relatively few viruses and many data are available. The major difficulties occur with the tailed phages where classification critically depends on good electron microscopy, special features such as unusual bases, and comparative serology and nucleic acid hybridization. *Ackermann* et al. [1978] analyzed the criteria used for describing and classifying phages. Major difficulties with the various criteria remain as follows.

Morphology. Unrelated phages may have identical morphology. Reported dimensions depend on the calibration of the electron microscope and are often unreliable.

Physicochemical Properties. The usefulness of such properties with the tailed phages is limited because of (i) parallelisms between capsid size and molecular weight of the DNA; the relative DNA content is mostly around 50%; (ii) parallelisms between GC content of phage and host cell; (iii) the size of tailed phages varies over a wide range without clear size classes; and (iv) particle density in CsCl is usually close to 1.5 g/cm^3.

Serology. Serological tests are usually carried out by neutralization of infectivity, which involves only the tip of the tail.

Resistance Tests. Tests involving resistance to various agents are performed in different laboratories under such variable conditions as to be almost meaningless.

Nucleic Acid Hybridization. At present the main problem is lack of data. Data are currently available for some 85 phages, mainly infecting enterobacteria. The threshold of relatedness is ill-defined. Several methods and many modifications of procedure have been reported. In general the data confirm relationships established by morphology, serology and other procedures.

Dr. *Ackermann* has received from the Natural Sciences and Engineering Research Council of Canada funds for a phage reference center. The grant is for 3 years and will provide for a technician and some material ($ 23,000 per year). The funding organization has suggested that the center be named after *Félix d'Hérelle*. The center will be taxonomically oriented and devoted to the preservation of type viruses. The center plans to: (i) collect and preserve type viruses of phage species, certain typing and teaching phages, and phages specific for resistance plasmids; (ii) check the identity of these viruses by electron microscopy or otherwise; (iii) provide on request reference phages and electron microscopical expertise for the identification of new phages; and (iv) publish a catalogue.

If adequately supported by bacterial virologists, this center should develop into a most useful tool for bacterial virus taxonomy.

B. Fungal Virus Subcommittee

The Chairman of this subcommittee, Dr. *M. Hollings*, retired at the end of 1980, and the chairman-elect *K.W. Buck* became Acting Chairman for the period January–August, 1981.

The main thrusts in mycovirus research at the present time are in four areas. *(i) Fungal killer systems:* There is now good evidence that killer proteins (fungal analogues of bacteriocins) are encoded by satellite double-stranded (ds) RNAs, the encapsidation and replication of which are dependent on dsRNA helper viruses. *(ii) Viruses of plant pathogenic fungi:* Evidence for the association of dsRNA with transmissible hypovirulence in *Endothia parasitica*, the causative agent of chestnut blight, has renewed interest in the effect of dsRNA and viruses on other plant pathogenic fungi *(iii) Viruses of edible fungi:* Work on diseases of mushrooms is continuing. In view of reports of virus-like particles in healthy, as well as diseased, mushrooms, much more work is required in this area. *(iv) Viruses of human pathogenic fungi:* A start in this area has been made in a number of laboratories. Although none of the above areas of research is directed specifically towards virus taxonomy, nevertheless useful taxonomic data are being accumulated. A large number of variants of some viruses obtained from different fungal isolates has been discovered, and this has helped to solve some taxonomic problems, while creating others.

Many mycoviruses have no obvious biological effects on their hosts and there is usually no reliable method of infecting fungi with their viruses. For these reasons the in vitro properties of the virus particles are of particular taxonomic importance. Many of the viruses infecting fungi have small isometric particles and contain dsRNA. Sufficient information is now available for many of these to allow the delineation of three families and of genera within them. Although these families have not yet been considered by ICTV, I have included them in the virus descriptions as proposed families to indicate the way in which fungal virus taxonomy is developing.

C. Invertebrate Virus Subcommittee

Arrangements have been made to ensure that the membership of vertebrate virus study groups includes at least one invertebrate virologist.

Two new families of viruses infecting invertebrates were proposed through the Coordination Subcommittee and were approved. The family ***Nodaviridae*** encompasses a group of small icosahedral viruses with divided single-stranded (ss)RNA genomes. The type genus is *Nodavirus* and the type species is Nodamura virus. The second family, which does not yet have an approved name, encompasses a group of small icosahedral ssRNA viruses with $T = 4$ symmetry. The type species for the new family is the *Nudaurelia* β virus. These two new families are now the responsibility of the Invertebrate Virus Subcommittee.

Proposals will be brought forward in the near future for another new family to include viruses whose genomes consist of ds, circular, polydisperse DNA. These viruses are associated with parasitic Braconid wasps.

D. Plant Virus Subcommittee

One new plant virus group was approved with carnation ringspot virus as the type member. The name ***Dianthovirus*** was approved for this group, which includes several viruses with two molecules of positive-sense ssRNA and polyhedral particles 31–34 nm in diameter. The name ***Sobemovirus*** was approved for the Southern bean mosaic virus group.

In Australia several plant viruses have recently been discovered in which the genome consists of two pieces of RNA. One is a linear ssRNA with a molecular weight of 1.5×10^6 while the other is a circular ss viroid-like RNA with a molecular weight of 1.2×10^5. This group is typified by velvet tobacco mottle virus. Although not yet approved by ICTV, a brief description is included in this report as a possible group.

The chairman of the subcommittee published an account of plant virus taxonomy – including a discussion of the historical background, current status, and future outlook [*Francki*, 1981].

E. Vertebrate Virus Subcommittee

As in previous terms, the Vertebrate Virus Subcommittee has been the most active of the host-oriented subcommittees of ICTV. However, many difficult problems remain for the future. The most innovative proposals approved by ICTV concerned the ***Adenoviridae.*** A definition for species within this family was approved. 87 species were approved in the genus *Mastadenovirus* and 14 in the genus *Aviadenovirus*. The proposed names for these species were also approved. They are constructed from the first three letters of the host's generic name followed by a number, the numbers being already in use by virologists. The exceptions are the human adenovirus series, where the letter 'h' followed by a number constitutes the species name. Under new rule No. 18 concerning species, approval of these species proposals by ICTV is provisional. Final approval can be given in 1984. Human adenoviruses were divided into five subgenera (A–E).

Proposals that the caliciviruses be excluded from the family ***Picornaviridae,*** and that a new family to be called the ***Caliciviridae*** be established, were approved.

Proposals for a substructure within the large family ***Bunyaviridae*** were approved. In addition to the established genus *Bunyavirus*, there are now three new genera with the following names (type species in parentheses) *Nairovirus* (Crimean-Congo hemorrhagic fever virus); *Phlebovirus* (sandfly fever Sicilian virus); *Uukuvirus* (Uukuniemi virus).

Proposals concerning the established family ***Parvoviridae*** were put forward through the Coordination Subcommittee. Two taxonomic proposals were approved: (i) two additional viruses were accepted as members of the genus *Parvovirus;* (ii) the adeno-associated viruses were grouped in a genus to be called *Dependovirus*. This last decision resolved a problem of many years' standing.

F. Coordination Subcommittee

The study groups concerned with ***Poxviridae, Reoviridae, Rhabdoviridae*** and ***Parvoviridae*** have worked closely with Dr. *Murphy* and the Vertebrate Virus Subcommittee during the period 1978–1981, as they did during the previous term. Although formally under the Coordination Subcommittee, the work of these study groups lies primarily in the vertebrate field. Some proposals put forward through the Coordination Subcommittee have already been noted under the Invertebrate and Vertebrate Virus Subcommittees.

The ***Iridoviridae*** study group led by Dr. *D. B. Willis* put forward ten taxonomic proposals which were approved by ICTV at Strasbourg. As a result, there are now five genera within this family (type species in parentheses): (i) *Iridovirus*, reserved for the 'small' iridoviruses (*Tipula* iridescent virus); (ii) *Chloriridovirus*, to include the large (180 nm) insect iridescent viruses (mosquito iridescent virus type 3); (iii) *Ranavirus*, to include a group of iridescent viruses infecting amphibians (frog virus 3); (iv) African swine fever virus genus (African swine fever virus); this is the only member of the ***Iridoviridae*** infecting mammals; and (v) Lymphocystis disease virus genus (the proposed type species is the lymphocystis disease virus of fish).

During the past term a new study group was set up under Dr. *P. Dobos* to consider viruses with genomes consisting of bisegmented dsRNA which infect vertebrates or invertebrates. No proposals from this study group have yet been approved by ICTV. However, I include in this report a possible family for these viruses that is being considered by the study group.

G. Standing Subcommittee for Virus Data (formerly the Code and Data Subcommittee)

When ICNV was first set up, a Cryptogram Subcommittee was formed to investigate the form and usefulness of the cryptogram in virus taxonomy [*Wildy*, 1971]. Later, this committee was given wider terms of reference and renamed the Code and Data Subcommittee [*Fenner*, 1976]. The cryptogram served a useful function in the early development of viral taxonomy, although it was never adopted officially by ICTV. In 1977 the Executive Committee decided to cease using the cryptogram in ICTV publications. The main reason for this decision was the inflexibility of the cryptogram in the light of important new characteristics of viruses that continue to be discovered.

Meantime, over the past 6 years under the chairmanship of *J. G. Atherton* in Brisbane, the Code and Data Subcommittee has evolved to have a substantially different kind of function than the other subcommittees of ICTV. The main work of the subcommittee became directed towards the development of a computer-based virus data storage and retrieval system applicable to all viruses. The committee made substantial progress towards the development of 'a code for the description of virus characters'. Such a code is essential for the computer handling of virus data. Among the advantages of the code are that it provides for systematic and uniform storage of data and allows for flexibility of data output. In addition, it provides inherent definition of terms to ensure standard description of characters.

The code is constructed in several sections, each with subcategories, and is written in questionnaire format to facilitate operator-computer interaction. The database management system in use is System 1022 T. M. (Software House, Cambridge, Mass.) on a DEC PDP KL-1090 computer. This is a highly efficient system, enabling interaction with visual display or hard-copy terminals. It has fast retrieval and updating facilities and provides for simple programming for output of data in any desired format. Input may be acquired from hard-copy or from specifically formated magnetic tape or disc.

Because there are not many virologists especially competent to take part in the development of such a project, increased stability for the membership of this committee became essential. Such a major program is not readily portable. Therefore, a permanent geographical base was needed. After considering proposals prepared by Dr. *J. G. Atherton* and Dr. *P. A. Bachmann*, the Executive Committee of ICTV at Strasbourg resolved as follows:

(i) That the Code and Data Subcommittee be replaced by a 'Standing Subcommittee for Virus Data'; such a committee to be appointed by the Executive Committee of ICTV for 3 years with no limit on reappointment. The chairman shall be appointed by the Executive Committee for a period of not more than two 3-year terms consecutively. A chairman will be eligible for reappointment after a gap of 3 years.

(ii) That the Standing Subcommittee be composed of persons directly concerned in handling virus data, with the power to add other persons as appropriate from time to time with the approval of the Executive Committee.

(iii) That the World Data Centre for Micro-organisms (Brisbane) be reaffirmed as the center for storage and retrieval of data relating to all viruses.

(iv) That the WHO Collaborating Centre for Collection and Evaluation of Data on Comparative Virology (Munich) be recognized as a center for storage and retrieval of data relating to viruses affecting vertebrates.

(v) That additional centers for other virus host groups be recognized from time to time.

(vi) That the Standing Subcommittee be asked to consider the means whereby data may be freely exchanged between recognized centers at minimal cost.

In view of the developments outlined above, the Executive Committee at Strasbourg resolved as follows: 'That ICTV subcommittees should send directly to the Data Centre in Brisbane authenticated basic data on viruses as they become available; and that such data should be reviewed and updated at the same time as the data for the ICTV report.'

V. Adequate Descriptions for New Virus Isolates

Published descriptions of new virus isolates are sometimes totally inadequate from the taxonomic point of view. This means that it may be impossible to determine whether the isolate is a newly described virus, a strain of a known virus, or an isolate indistinguishable from a previously described virus. This is a continuing problem that affects the work of all the host-oriented subcommittees to varying extents. Guidelines for the adequate description of viruses have been published under the guidance of the subcommittees for bacterial viruses [*Ackermann* et al., 1978] and for plant viruses [*Hamilton* et al., 1981]. It is to be hoped that the other host-oriented subcommittees will follow suit and that virologists in general will use the guidelines effectively.

VI. Standardization of Terms Used in Virology

From time to time it has been suggested that ICTV should become involved in providing a formal standardization of terms used in virology. Up to the present the Executive Committee has considered this to be an inappropriate task for ICTV. Nevertheless, a problem exists. When an ICTV report is prepared, the President frequently has the choice of using a variety of synonymous terms in the text or making a decision to standardize on a particular term. The second course is certainly preferable. To assist the President, as an interim measure, the Executive Committee designated the Coordination Subcommittee to advise the President on matters of terminology.

In the near future it may well prove useful for ICTV to play a more positive role in the standardization of virological terminology. This might be done by a continuing survey of terms used in the literature, with a list of terms and definitions (either approved or preferred) being published in each report of ICTV.

The Initiation and Processing of New Taxonomic Proposals

Contents

Over recent years the Executive Committee of ICTV has evolved procedures and rules to facilitate the processing and assessment of new taxonomic proposals for viruses. In this section I will summarize the present position, to assist virologists wishing to make a contribution to the work of ICTV.

I. Initiation of New Proposals

The key units in the organization of ICTV are the host-oriented subcommittees. Most of these subcommittees set up study groups consisting of working virologists who are expert with respect to a particular family of viruses. New taxonomic proposals are usually initiated by these study groups, and less commonly by the subcommittees themselves.

Subcommittees are not bound by the proposals of their study groups. They may decline the proposals or make changes in them with or without reference back to the study group, although normally the study group would be consulted as far as possible.

It should be emphasized that, apart from the formal organization, it is perfectly in order for any individual virologist to initiate a new taxonomic proposal. Any such proposal should be in the format outlined below, and should be sent to the Chairman of the appropriate subcommittee for consideration.

II. Publication of New Proposals

Some new proposals pass through the ICTV organization as outlined in the next section and are approved by ICTV without any prior publication. Such proposals then appear first in an official ICTV triennial report. Other proposals are published at an earlier stage, usually in *Intervirology*, which is the Journal of the Virology Division of the International Union of Microbiological Societies.

These reports may be enlarged presentations of taxonomic proposals being formally submitted by ICTV study groups. Two recent examples of this sort concern the ***Caliciviridae*** family [*Schaffer* et al., 1980] and the ***Bunyaviridae*** [*Bishop* et al., 1980]. Others may be taxonomic proposals prepared by groups or organizations not formally associated with ICTV, for example, proposals on simian virus nomenclature by the Simian Virus Working Team of the WHO/FAO Programme on Comparative Virology [*Kalter* et al., 1980] and a proposed classification of viruses within the *Alphavirus* genus by a subcommittee of the American Committee on Arthropod-borne Viruses [*Calisher* et al., 1980]. Other published proposals may come from individual virologists, for example, suggestions for the classification of helical plant viruses [*Veerisetty*, 1979].

Such publications allow individual virologists to scrutinize proposals and to make their views known to the appropriate ICTV subcommittee. It should be emphasized, however, that publication in itself does not give the proposals any status as far as ICTV is concerned.

III. Processing of New Proposals

A taxonomic proposal originating in a study group is forwarded to the appropriate subcommittee. If it is approved by the subcommittee, the proposal is then considered by the Executive Committee of ICTV. The Executive Committee may approve a proposal, decline to approve, or send it back to the subcommittee with suggested changes. The Executive Committee does not make changes in a taxonomic proposal without referral back to the subcommittee. A significant proportion of the taxonomic proposals coming before the Executive Committee for the first time are not approved, for a variety of reasons.

Proposals approved by the Executive Committee go forward every 3 years to the plenary meeting of the full ICTV membership for ratification. At this final stage, very few proposals are not approved.

IV. Timing of Events in the Period 1981–1984

The Executive Committee of ICTV will hold its mid-term meeting in April 1983 in London. At this meeting it will consider any taxonomic proposals put before it. There is no deadline before the meeting for receipt of proposals.

Subcommittee chairmen can send proposals to the Secretary, *J. Maurin*, for circulation to members before meeting; he can mail them directly to members or they can be tabled at the meeting.

Following the mid-term meeting any further new taxonomic proposals must be in the hands of the Secretary, *J. Maurin*, by May 1, 1984. This is to allow time for circulation of all new proposals to ICTV members for consideration before the ICTV meetings to be held in conjunction with the Sixth International Congress of Virology in Sendai, Japan, in August 1984.

In August 1984 the Executive Committee could consider and approve new taxonomic proposals received after May 1984, but such proposals would have to wait a further 3 years for ratification by ICTV in 1987.

It can be seen from the above timetable that there are advantages in submitting new taxonomic proposals in time for the mid-term meeting of the Executive Committee in April 1983. This gives an opportunity for any proposals that have been declined to be modified and resubmitted by May 1984.

V. Standard Format for Presenting New Taxonomic Proposals

Study group chairmen and subcommittee chairmen should use the following guidelines and format in preparing new taxonomic proposals.

Guidelines:

1. Each individual taxonomic proposal should be submitted as a separate item (and not mixed with explanatory or historical detail). For example, a proposal to form a new genus is separate from any proposed name for that genus and from the designation of a type species for the genus.

2. Attention is drawn to *new rule No.22*, which requires that approval of a new family must be linked with approval of a type genus and that approval of a new genus must be linked with approval of a type species.
3. Each proposal should contain the information in the following format:

Date ....

From the ... ... Subcommittee or Study Group

Taxonomic Proposal No.:

1. *Proposal:* The taxonomic proposal in its essence, in a form suitable for presentation to ICTV for voting.
2. *Purpose:* A summary of the reasons for the proposal, with any explanatory and historical notes.
3. A summary of the new taxonomic situation within the family, group or genus (e.g., for a new genus – 'The family would now consist of the following genera: ... ...').
4. Derivation of any names proposed.
5. New literature references, if appropriate.

The Rules of Nomenclature of Viruses

The rules governing the nomenclature of viruses are not changed by ICTV without very good reason. Over the past 15 years a sound framework has been developed for virus families and to a lesser extent for genera, particularly for viruses infecting vertebrates. It became apparent about 3–4 years ago that the stage was being reached for many families where further taxonomic progress would depend on the delineation of virus species and the allocation of official international names to them.

However, the three rules governing species had remained essentially unchanged since they were first agreed upon in 1966. The very general wording of these rules made them of little use in guiding virologists who were attempting to delineate and name virus species.

The issues were discussed in depth by the Executive Committee in April 1980. As a result, it was recommended that old rules 11, 12, and 13 dealing with species be deleted and replaced with eight new rules (numbers 12–18 below). A further proposed new rule (number 22 below) was designed to link the approval of a new family to an approved type genus and the approval of a new genus to an approved type species. These proposed changes were circulated to all members of ICTV more than 1 year before the meeting, as required by the policy of ICTV. At the Plenary Meeting in Strasbourg, ICTV approved all the rule changes with only one minor change in wording.

In addition, the Executive Committee approved the set of seven 'guidelines' which follow the rules below. These guidelines are intended to supplement the rules and assist virologists to formulate species proposals which will meet with general approval and which will further facilitate the development of a reasonably uniform taxonomy and nomenclature for all viruses.

Rule 1 – The code of bacterial nomenclature shall not be applied to viruses.

Rule 2 – Nomenclature shall be international.

Rule 3 – Nomenclature shall be universally applied to all viruses.

Rule 4 – An effort will be made towards a latinized nomenclature.

Rule 5 – Existing latinized names shall be retained whenever feasible.

Rule 6 – The law of priority shall not be observed.

Rule 7 – Sigla may be accepted as names of viruses or virus groups, provided that they are meaningful to workers in the field and are recommended by international virus study groups.

Rule 8 – No person's name shall be used.

Rule 9 – Names should have international meaning.

Rule 10 – The rules of orthography of names and epithets are listed in Chapter 3, Section 6, of the proposed international code of nomenclature of viruses [Appendix D; Minutes of 1966 (Moscow) meeting].

New Rule 11 – A virus species is a concept

that will normally be represented by a cluster of strains from a variety of sources, or a population of strains from a particular source, which have in common a set or pattern of correlating stable properties that separates the cluster from other clusters of strains.

New Rule 12 – The genus name and species epithet, together with the strain designation, must give an unambiguous identification of the virus.

New Rule 13 – The species epithet must follow the genus name and be placed before the designation of strain, variant or serotype.

New Rule 14 – A species epithet should consist of a single word or, if essential, a hyphenated word. The word may be followed by numbers or letters.

New Rule 15 – Numbers, letters, or combinations thereof may be used as an official species epithet where such numbers or letters already have wide usage for a particular virus.

New Rule 16 – Newly designated serial numbers, letters or combinations thereof are not acceptable alone as species epithets.

New Rule 17 – Artificially created laboratory hybrids between different viruses will not be given taxonomic consideration.

New Rule 18 – Approval by ICTV of newly proposed species, species names and type species will proceed in two stages. In the first stage, provisional approval may be given. Provisionally approved proposals will be published in an ICTV report. In the second stage, after a 3-year waiting period, the proposals may receive the definitive approval of ICTV.

Rule 19 – The genus is a group of species sharing certain common characters.

Rule 20 – The ending of the name of a viral genus is '. . . *virus*'.

Rule 21 – A family is a group of genera with common characters, and the ending of the name of a viral family is ***'. . . viridae'.***

New Rule 22 – Approval of a new family must be linked to approval of a type genus; approval of a new genus must be linked to approval of a type species.

Guidelines for the Delineation and Naming of Species

1. Criteria for delineating species may vary in different families of viruses.
2. Wherever possible, duplication of an already approved virus species name should be avoided.
3. When a change in the type species is desirable, this should be put forward to ICTV in the standard format for a taxonomic proposal.
4. Subscripts, superscripts, hyphens, oblique bars, or Greek letters should be avoided in future virus nomenclature.
5. When designating new virus names, study groups should recognize national

sensitivities with regard to language. If a name is universally used by virologists (those who publish in scientific journals), that name or a derivative of it should be used regardless of national origin. If different names are used by virologists of different national origin, the study group should evaluate relative international usage and recommend the name that will be acceptable to the majority and which will not be offensive in any language.

6. ICTV is not concerned with the classification and naming of strains, variants or serotypes. This is the responsibility of specialist groups.
7. Virus taxonomy at its present stage has no evolutionary or phylogenetic implications.

The Viruses

Presentation

This report contains a listing of the virus taxa approved by ICTV between 1970 and 1981. Descriptions of the important characteristics of these taxa are provided, together with a list of members and selected references giving a guide to recent literature. The detailed information has been provided from the work of the subcommittees of ICTV and of their various study groups, and from individual virologists.

Names for Viruses, Genera, and Families

In the formal descriptions (pp. 42–178) the family, subfamily, genus and species names approved by ICTV are listed under 'International name'. All names of taxa approved by ICTV are printed in italic type, as was agreed by the Executive Committee at Strasbourg.

Names that have not been officially approved are printed in standard type face. The heading 'English vernacular name' is used, even though for a few viruses a name in some other language has been adopted into English usage. Where there is a widely used vernacular synonym, this is included within parentheses. In the virus diagrams, approved names for all taxa are in bold type. For those plant viruses that have been included in the CMI/AAB Descriptions of Plant Viruses the description number is given in parentheses following the name.

Main Characteristics

The 'Main characteristics' section has been further expanded for most taxa. The order of listing of data is standardized for ease of reference. As would be expected, the amount of relevant information available varies quite widely for different families, genera and groups. Since all known plant viruses can be transmitted by grafting and vegetative propagation, these two methods of transmission have been omitted in the descriptions.

List of Members

The lists of members for genera and groups have been updated. In these lists the word 'virus' has been omitted for the sake of brevity, unless it forms part of a single word in the name or unless the plural 'viruses' is required. Three categories of members have been defined as follows:

Other members: Those viruses, besides the type member, which definitely belong in the family, genus or group.

Probable members: Those viruses for which information known to study group members strongly indicates affiliation with the family, genus or group.

Possible members: Viruses or isolates for which taxonomically useful data must be regarded as more tenuous.

To assist readers, fairly extensive lists of names have been included for many of the taxa. It should be remembered, however, that these lists may contain described and named isolates which, on further examination, will be shown to be closely related strains or even indistinguishable isolates of a single virus.

Arrangement of the Approved Families and Groups

54 families and groups of viruses have now been approved by ICTV. Since a taxonomic structure above the level of family has not yet been developed, any sequences of listing must

be arbitrary. Many virologists consider the kind, and strandedness, of the nucleic acid making up the viral genome and the presence or absence of a lipoprotein envelope to be basically important virus properties. Using these three properties, the 54 approved families and groups (and several proposed or tentative families and groups) are described in the order listed in table II. There are no known ssDNA viruses with envelopes, so these three virus properties give rise to seven clusters of families and groups.

Within two of these clusters, the families can be usefully arranged on other criteria as follows: (i) for the enveloped ssRNA viruses, on the basis of genome strategy [*Baltimore*, 1971; *Cooper*, 1974]; and (ii) for the non-enveloped ssRNA viruses infecting primarily plants, on the basis of particle morphology and on the number of pieces of RNA comprising the genome (table II). In addition, to save repetition, a general description is given to cover the three families of tailed phages. These arrangements remain unchanged from the Third Report. They are not intended to anticipate higher taxa, a subject that has not yet been considered by ICTV.

Possible or Tentative Families and Groups

Brief descriptions for several possible or tentative families and groups are included in appropriate positions among the approved taxa. These are: three proposed families of dsRNA viruses infecting fungi, the bisegmented dsRNA viruses infecting vertebrates and invertebrates, and the velvet tobacco mottle virus group infecting plants. A list of important but unclassified viruses or virus-like agents is given at the end of the report.

Index

Following the virus descriptions, there is a single index containing all the virus names used in the text. Family, genus and group names approved by ICTV are given in italics. In addition to the main index, page numbers for the approved families and groups are given in table II and in the four pages of line drawings for the vertebrate, invertebrate, plant, and bacterial viruses.

Glossary of Some Abbreviations and Virological Terms as Used in the Virus Descriptions

(Note: These terms are approved by the Coordination Subcommittee of ICTV for use in this Report but have no official status.)

(i) Abbreviations

CF = complement fixing
CPE = cytopathic effect
D = diffusion coefficient
DI = defective interfering
ds = double-stranded
HI = hemagglutination inhibition
MW = molecular weight
RF = replicative form
RI = replicative intermediate
RNP = ribonucleoprotein
ss = single-stranded

(ii) RNA Replicases, Transcriptases and Polymerases:

In the synthesis of viral RNA, the term polymerase has been replaced in general by two somewhat more specific terms: RNA replicase and RNA transcriptase. The term replicase refers to the enzyme responsible for synthesis of the progeny genome RNA of whatever polarity. The term transcriptase has become associated with the enzyme involved in messenger RNA synthesis, most recently with those polymerases which are virion-associated. However, it should be borne in mind that for some viruses it has yet to be established whether or not the replicase and transcriptase activities reflect distinct enzymes rather than alternate activities of a single enzyme. Confusion also arises in the case of the small positive-sense RNA viruses where the term replicase (e.g., Qβ replicase) has been used for the enzyme capable both of transcribing the genome into messenger RNA via an intermediate negative-sense strand and of synthesizing the genome strand from the same template. In the text, the term replicase will be restricted as far as possible to the enzyme synthesizing progeny viral strands of either polarity. The term transcriptase is restricted to those RNA polymerases that are virion-associated and synthesize mRNA. The generalized term RNA polymerase (i.e., RNA-dependent RNA polymerase) is applied where no distinction between replication and transcription enzymes can be drawn (e.g., Qβ, R17, poliovirus and many plant viruses).

(iii) Other Definitions:

Enveloped: possessing an outer (bounding) lipoprotein bilayer membrane.

Negative-sense strand: (= minus strand); for RNA or DNA, the strand with a base sequence complementary to the positive-sense strand.

Positive-sense strand: (= plus strand, message strand); for RNA, the strand that contains the coding triplets which can be translated on ribosomes. For DNA, the strand that contains the same base sequence as the mRNA. However, in some dsDNA viruses mRNAs are transcribed from both strands and the transcribed regions may overlap. For such viruses this definition is inappropriate.

Pseudotypes	Enveloped virus particles in which the envelope is derived from one virus and the internal constituents from another.
Reverse transcriptase:	Virus-encoded RNA-dependent DNA polymerase found as part of the virus particle in the ***Retroviridae.***
Surface projections:	(= spikes, peplomers, knobs); morphological features, usually consisting of glycoproteins, that protrude from the lipoprotein envelope of many enveloped viruses.
Virion:	Morphologically complete virus particle.
Viroplasm:	(= virus factory, virus inclusion, X-body); a modified region within the infected cell in which virus replication occurs, or is thought to occur.

Virus Diagrams

The following four pages provide line drawings for the virus families and groups arranged according to the major host (vertebrate, invertebrate, plant or bacterial).

All the diagrams have been drawn to the same scale. They give a good indication of the shapes and the relative sizes of the viruses, but they cannot be taken as definitive for several reasons: (i) Different viruses within a family may vary somewhat in size and shape. For some families the dimensions of a well-characterized member were taken. For others an 'average' for many members has been used (e.g., with the tailed phages). Where differences are large within a family, two drawings are provided (e.g., ***Inoviridae***). (ii) Dimensions of some viruses are difficult to determine. (iii) Some viruses, particularly the larger enveloped ones, may be pleomorphic.

Most of the smaller viruses are given in outline only, with an indication of icosahedral structure where appropriate. The large viruses are given schematically in surface outline, in section, or both, as seems most appropriate to display major morphological characteristics.

For each drawing the family or group name is given along with the relevant page number for a description of the family. To aid recognition a well-known member of the family is named, but the dimensions and shape used for the drawing may not be exactly those of the virus named. The outline for *Geminivirus* is based on the model favored by *Hatta and Francki* [1979] for chloris striate mosaic virus.

I wish to thank Dr. *H.-W. Ackermann* for providing the outline drawings of bacterial viruses, and Mrs. *J. Keeling* for preparing the drawings of viruses for the other host groups.

THE FAMILIES OF VIRUSES INFECTING BACTERIA

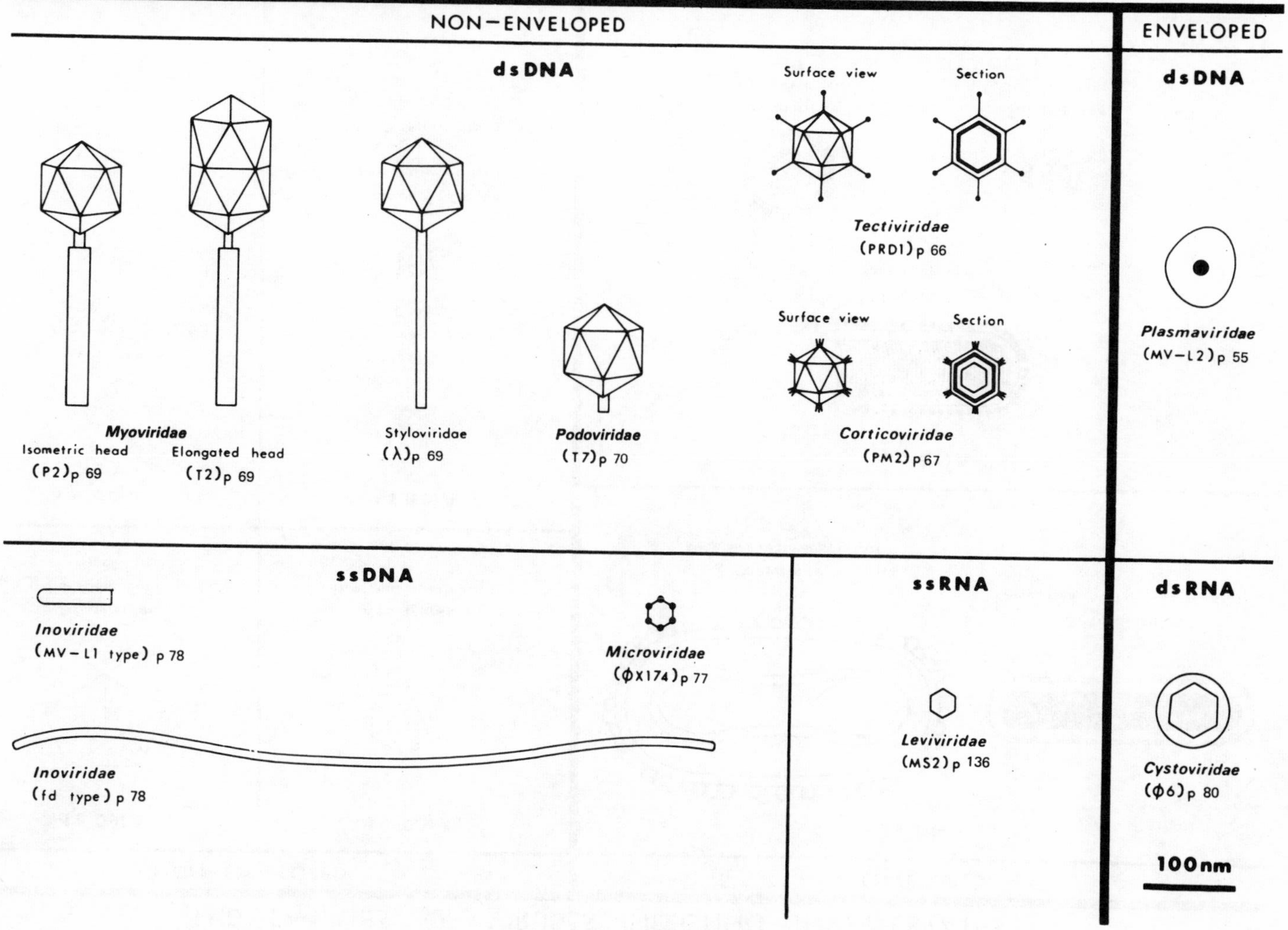

NON-ENVELOPED

dsDNA

Myoviridae

Isometric head
(P2) p 69

Elongated head
(T2) p 69

Styloviridae
(λ) p 69

Podoviridae
(T7) p 70

Surface view
Section

Tectiviridae
(PRD1) p 66

Surface view
Section

Corticoviridae
(PM2) p 67

ssDNA

Inoviridae
(MV-L1 type) p 78

Microviridae
(φX174) p 77

Inoviridae
(fd type) p 78

ssRNA

Leviviridae
(MS2) p 136

ENVELOPED

dsDNA

Plasmaviridae
(MV-L2) p 55

dsRNA

Cystoviridae
(φ6) p 80

100nm

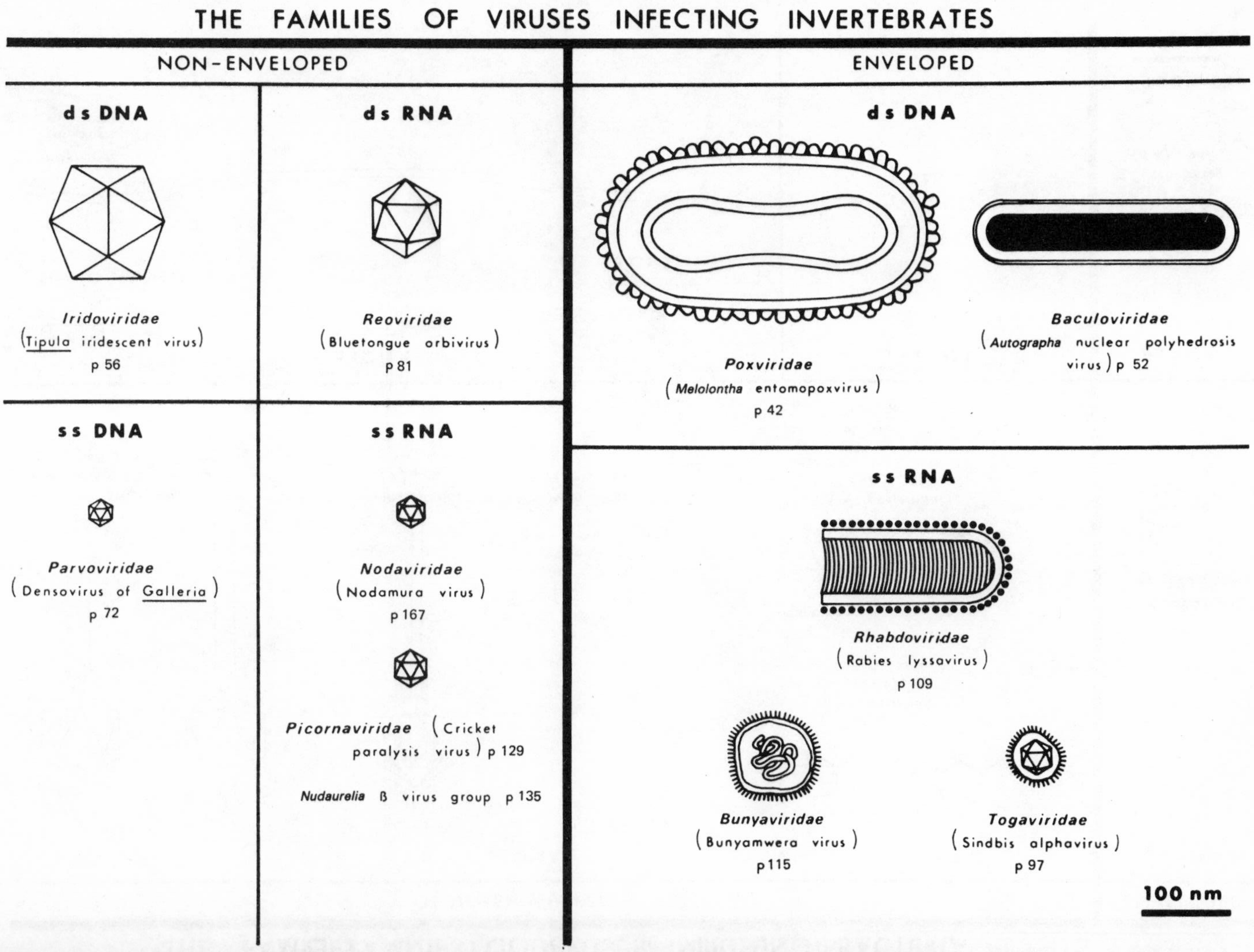

THE FAMILIES OF VIRUSES INFECTING INVERTEBRATES

NON-ENVELOPED

ds DNA

Iridoviridae
(Tipula iridescent virus)
p 56

ds RNA

Reoviridae
(Bluetongue orbivirus)
p 81

ss DNA

Parvoviridae
(Densovirus of Galleria)
p 72

ss RNA

Nodaviridae
(Nodamura virus)
p 167

Picornaviridae (Cricket paralysis virus) p 129

Nudaurelia ß virus group p 135

ENVELOPED

ds DNA

Poxviridae
(*Melolontha* entomopoxvirus)
p 42

Baculoviridae
(*Autographa* nuclear polyhedrosis virus) p 52

ss RNA

Rhabdoviridae
(Rabies lyssavirus)
p 109

Bunyaviridae
(Bunyamwera virus)
p 115

Togaviridae
(Sindbis alphavirus)
p 97

100 nm

THE FAMILIES AND GROUPS OF VIRUSES INFECTING PLANTS

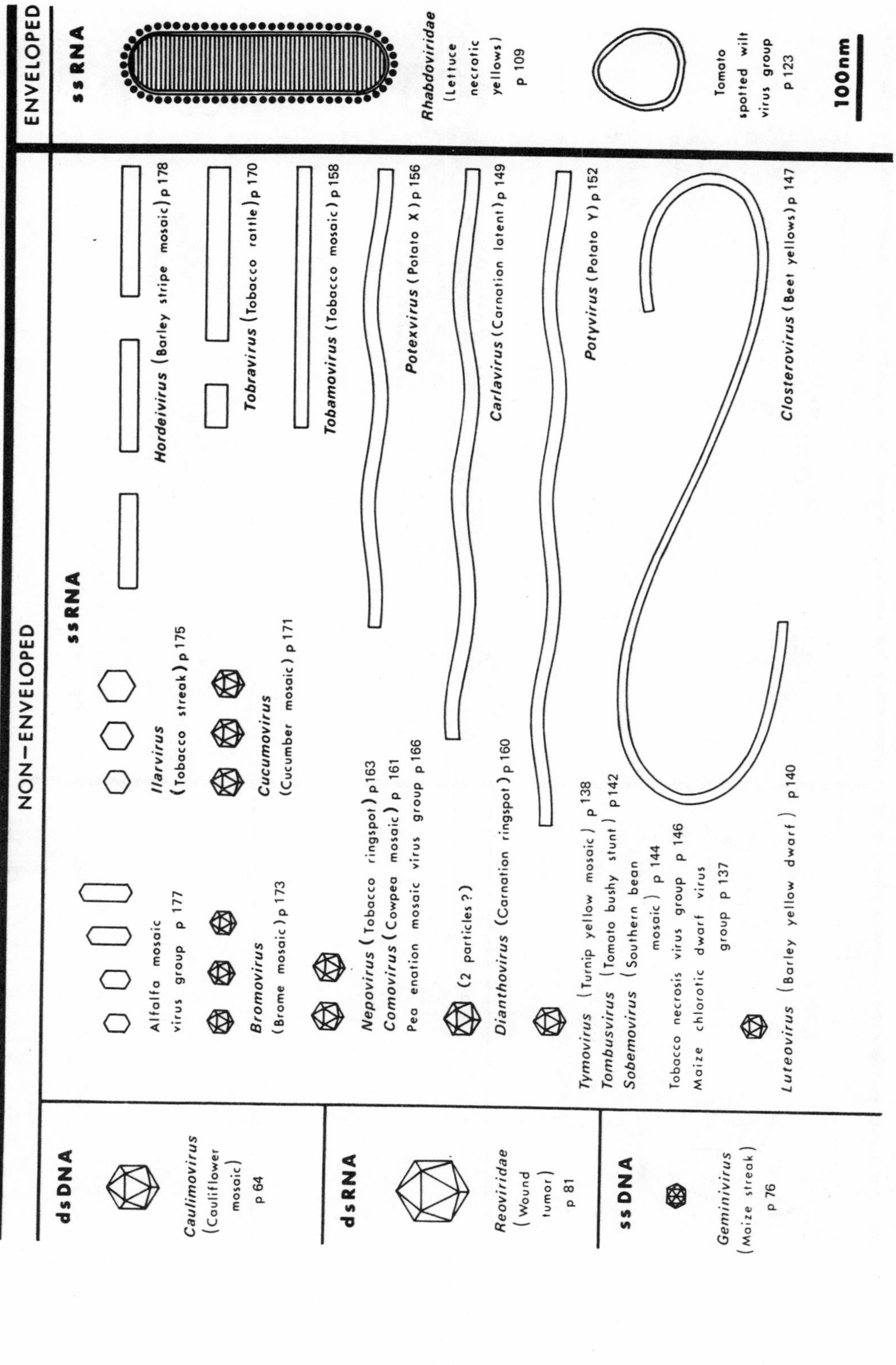

NON-ENVELOPED

dsDNA

Caulimovirus (Cauliflower mosaic) p 64

dsRNA

Reoviridae (Wound tumor) p 81

ssDNA

Geminivirus (Maize streak) p 76

ssRNA

Alfalfa mosaic virus group p 177

Ilarvirus (Tobacco streak) p 175

Hordeivirus (Barley stripe mosaic) p 178

Bromovirus (Brome mosaic) p 173

Cucumovirus (Cucumber mosaic) p 171

Tobravirus (Tobacco rattle) p 170

Tobamovirus (Tobacco mosaic) p 158

Nepovirus (Tobacco ringspot) p 163

Comovirus (Cowpea mosaic) p 161

Pea enation mosaic virus group p 166

Potexvirus (Potato X) p 156

(2 particles ?)

Dianthovirus (Carnation ringspot) p 160

Carlavirus (Carnation latent) p 149

Potyvirus (Potato Y) p 152

Tymovirus (Turnip yellow mosaic) p 138

Tombusvirus (Tomato bushy stunt) p 142

Sobemovirus (Southern bean mosaic) p 144

Tobacco necrosis virus group p 146

Maize chlorotic dwarf virus group p 137

Luteovirus (Barley yellow dwarf) p 140

Closterovirus (Beet yellows) p 147

ENVELOPED

ssRNA

Rhabdoviridae (Lettuce necrotic yellows) p 109

Tomato spotted wilt virus group p 123

100nm

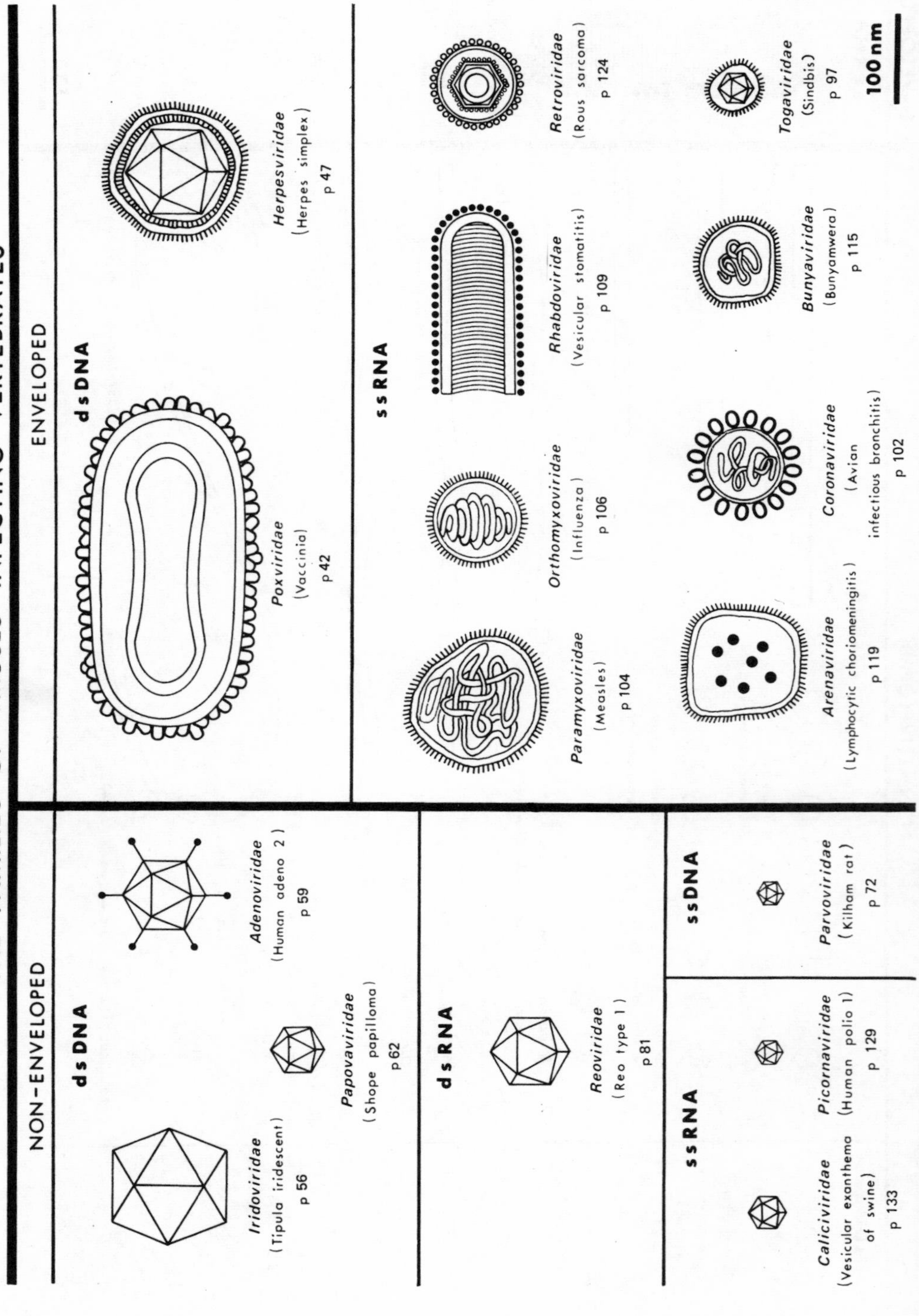

THE FAMILIES OF VIRUSES INFECTING VERTEBRATES

NON-ENVELOPED

dsDNA

Iridoviridae
(Tipula iridescent)
p 56

Papovaviridae
(Shope papilloma)
p 62

Adenoviridae
(Human adeno 2)
p 59

dsRNA

Reoviridae
(Reo type 1)
p 81

ssRNA

Caliciviridae
(Vesicular exanthema of swine)
p 133

Picornaviridae
(Human polio 1)
p 129

ssDNA

Parvoviridae
(Kilham rat)
p 72

ENVELOPED

dsDNA

Poxviridae
(Vaccinia)
p 42

Herpesviridae
(Herpes simplex)
p 47

ssRNA

Paramyxoviridae
(Measles)
p 104

Orthomyxoviridae
(Influenza)
p 106

Rhabdoviridae
(Vesicular stomatitis)
p 109

Retroviridae
(Rous sarcoma)
p 124

Arenaviridae
(Lymphocytic choriomeningitis)
p 119

Coronaviridae
(Avian infectious bronchitis)
p 102

Bunyaviridae
(Bunyamwera)
p 115

Togaviridae
(Sindbis)
p 97

100 nm

Table II. List of virus families and groups in order of presentation

Characterization		Families or groups	Kinds of host	Page No.
dsDNA, enveloped		***Poxviridae***	vertebrates, invertebrates	42
		Herpesviridae	vertebrates	47
		Baculoviridae	invertebrates	52
		Plasmaviridae	bacteria	55
dsDNA, nonenveloped		***Iridoviridae***[1]	vertebrates, invertebrates	56
		Adenoviridae	vertebrates	59
		Papovaviridae	vertebrates	62
		Caulimovirus	plants	64
		Tectiviridae	bacteria	66
		Corticoviridae	bacteria	67
	tailed phages	***Myoviridae***	bacteria	68
	tailed phages	Styloviridae	bacteria	69
	tailed phages	***Podoviridae***	bacteria	70
ssDNA, nonenveloped		***Parvoviridae***	vertebrates, invertebrates	72
		Geminivirus	plants	76
		Microviridae	bacteria	77
		Inoviridae	bacteria	78
dsRNA, enveloped		***Cystoviridae***	bacteria	80
dsRNA, nonenveloped		***Reoviridae***	vertebrates, invertebrates, plants	81
	possible family	Isometric dsRNA mycoviruses requiring only one RNA segment for replication	fungi	89
	possible family	Isometric dsRNA mycoviruses requiring two RNA segments for replication	fungi	91
	possible family	Isometric dsRNA mycoviruses requiring three RNA segments for replication	fungi	94
	possible family	Bisegmented dsRNA viruses	vertebrates, invertebrates	95

[1] Some members possess a cell membrane-derived envelope (see p. 56).

Table II. (continued)

Characterization		Families or groups	Kinds of host	Page No.
ssRNA, enveloped				
Genome strategy	*a. No DNA step*			
	(i) positive-sense genome	***Togaviridae***	vertebrates, invertebrates	97
		Coronaviridae	vertebrates	102
	(ii) negative-sense genome	***Paramyxoviridae***	vertebrates	104
		Orthomyxoviridae	vertebrates	106
		Rhabdoviridae	vertebrates, invertebrates, plants	109
		Bunyaviridae	vertebrates, invertebrates	115
		Arenaviridae	vertebrates	119
	(iii) not established	Tomato spotted wilt virus group	plants	123
	b. DNA step in replication cycle	***Retroviridae***	vertebrates	124
ssRNA, nonenveloped	*Monopartite genomes*			
	Isometric particles	***Picornaviridae***	vertebrates, invertebrates	129
		Caliciviridae	vertebrates	133
		Nudaurelia β virus group	invertebrates	135
		Leviviridae	bacteria	136
		Maize chlorotic dwarf virus group	plants	137
		Tymovirus	plants	138
		Luteovirus	plants	140
		Tombusvirus	plants	142
		Sobemovirus	plants	144
		Tobacco necrosis virus group	plants	146
	Rod-shaped particles	***Closterovirus***	plants	147
		Carlavirus	plants	149
		Potyvirus	plants	152
		Potexvirus	plants	156
		Tobamovirus	plants	158

Table II. (continued)

Characterization		Families or groups	Kinds of host	Page No.
	Bipartite genomes			
	Isometric particles	***Dianthovirus***	plants	160
		Comovirus	plants	161
		Nepovirus	plants	163
		Pea enation mosaic virus group	plants	166
		Nodaviridae	invertebrates	167
	Possible group	Velvet tobacco mottle virus group	plants	169
	Rod-shaped particles	***Tobravirus***	plants	170
	Tripartite genomes			
	Isometric particles	***Cucumovirus***	plants	171
		Bromovirus	plants	173
		Ilarvirus	plants	175
	Isometric and bacilliform particles	Alfalfa mosaic virus group	plants	177
	Rod-shaped particles	***Hordeivirus***	plants	178

References

Ackermann, H.W.; Audurier, A.; Berthiaume, L.; Jones, L.A.; Mayo, J.A.; Vidaver, A.K.: Guidelines for bacteriophage characterization. Adv. Virus Res. *23:* 1–24 (1978).

Baltimore, D.: Expression of animal virus genomes. Bact. Rev. *35:* 235–241 (1971).

Bishop, D.H.L.; Calisher, C.H.; Casals, J.; Chumakov, M.P.; Gaidamovich, S.Ya.; Hannoun, C.; Lvov, D.K.; Marshall, I.D.; Oker-Blom, N.; Pettersson, R.F.; Porterfield, J.S.; Russell, P.K.; Shope, R.E.; Westaway, E.G.: Bunyaviridae. Intervirology *14:* 125–143 (1980).

Calisher, C.H.; Shope, R.E.; Brandt, W.; Casals, J.; Karabatsos, N.; Murphy, F.A.; Tesh, R.B.; Wiebe, M.E.: Proposed antigenic classification of registered arboviruses. I. Togaviridae, *Alphavirus.* Intervirology *14:* 229–232 (1980).

Cooper, P.D.: Towards a more profound basis for the classification of viruses. Intervirology *4:* 317–319 (1974).

Fenner, F.: Classification and nomenclature of viruses. Second report of the International Committee on Taxonomy of Viruses. Intervirology *7:* 1–116 (1976).

Francki, R.I.B.: Plant virus taxonomy; in Kurstak, Handbook of plant virus infections and comparative diagnosis, pp. 1–16 (Elsevier/North Holland, Amsterdam 1981).

Hamilton, R.I.; Edwardson, J.R.; Francki, R.I.B.; Hsu, H.T.; Hull, R.; Koenig, R.; Milne, R.G.: Guidelines for the identification and characterization of plant viruses. J. gen. Virol. *54:* 223–241 (1981).

Hatta, T.; Francki, R.I.B.: The fine structure of chloris striate mosaic virus. Virology *92:* 428–435 (1979).

Kalter, S.S.; Ablashi, D.; Espana, C.; Heberling, R.L.; Hull, R.N.; Lennette, E.H.; Malherbe, H.H.; McConnell, S.; Yohn, D.S.: Simian virus nomenclature, 1980. Intervirology *13:* 317–330 (1980).

Matthews, R.E.F.: Classification and nomenclature of viruses. Third report of the International Committee on Taxonomy of Viruses. Intervirology *12:* 129–296 (1979).

Matthews, R.E.F.: The classification and nomenclature of viruses. Summary of results of meetings of the International Committee on Taxonomy of Viruses in Strasbourg, August 1981. Intervirology *16:* 53–60 (1981).

Schaffer, F.L.; Bachrach, H.L.; Brown, F.; Gillespie, J.H.; Burroughs, J.N.; Madin, S.H.; Madeley, C.R.; Povey, R.C.; Scott, F.; Smith, A.W.; Studdert, M.J.: Caliciviridae. Intervirology *14:* 1–6 (1980).

Veerisetty, V.: Suggestions for the classification and nomenclature of helical plant viruses. Intervirology *11:* 167–173 (1979).

Wildy, P.: Classification and nomenclature of viruses. First report of the International Committee on Nomenclature of Viruses. Monogr. Virol., vol. 5 (Karger, Basel 1971).

The Families and Groups

Taxonomic status	*English vernacular name*	*International name*
Family	Poxvirus group	***POXVIRIDAE***
Main characteristics	A. Properties of the Virus Particle Nucleic acid: Single molecule of dsDNA; MW in the range 85–240 × 10^6. G + C content of vertebrate poxviruses = 35–40%; of entomopoxviruses about 26%. Protein: More than 30 structural proteins and several viral enzymes concerned with nucleic acid synthesis and processing, including a DNA-dependent transcriptase. Lipid: About 4% by weight (vaccinia). Carbohydrate: About 3% by weight. Physicochemical properties: Infectivity is ether-resistant in some members, ether-sensitive in others. Morphology: Large, brick-shaped or ovoid virion, 300–450 nm × 170–260 nm, with external coat containing lipid and tubular or globular protein structures, enclosing one or two lateral bodies and a core, which contains the genome. B. Replication Multiplication occurs in cytoplasm, with type B (viral factory) and type A (cytoplasmic accumulation) inclusion bodies. Mature particles released from microvilli or by cellular disruption. Genetic recombination occurs within genera; nongenetic reactivation occurs both within and between genera of vertebrate poxviruses. Hemagglutinin, separate from the virion, is produced by viruses of *Orthopoxvirus* only. C. Biological Aspects Host range: Generally narrow in vertebrates or invertebrates. Transmission: Fomites. Airborne. Also by contact. Mechanical by arthropods is common.	
Subfamilies	Poxviruses of vertebrates Poxviruses of insects	***Chordopoxvirinae*** ***Entomopoxvirinae***

Subfamily	Poxviruses of vertebrates	***Chordopoxvirinae***
Main characteristics	Morphologically and chemically like other members of the family ***Poxviridae.*** About 10 major antigens in virion, one of which cross-reacts with most poxviruses of vertebrates. Extensive serological cross-reactivity within each of the genera of vertebrate poxviruses.	
Genera	Vaccinia subgroup Orf subgroup Fowlpox subgroup Sheep pox subgroup Myxoma subgroup Swinepox subgroup	*Orthopoxvirus* *Parapoxvirus* *Avipoxvirus* *Capripoxvirus* *Leporipoxvirus* *Suipoxvirus*

Genus	Vaccinia subgroup	*Orthopoxvirus*
Type species	Vaccinia virus	–

Taxonomic status	*English vernacular name*	*International name*
Main characteristics	A. Properties of the Virus Particle Single linear molecules of dsDNA with covalently closed ends. MW = 160×10^6. Infectivity of virions is ether-resistant. Different species undergo genetic recombination and exhibit extensive serological cross-reactivity and nucleic acid homology. Lipid of distinctive composition. B. Replication Intracellular virus and naturally released extracellular virus are antigenically different. A lipoprotein hemagglutinin is produced in infected cells and becomes incorporated into the modified cell membrane. It is probably found on the surface of extracellular virus. The A-type inclusions may or may not contain virions in different orthopoxviruses. C. Biological Aspects Host range: Of individual viruses is rather narrow; usually limited to a single animal host in nature.	
Other members	Buffalopox (buffalos) Camelpox (camels) Cowpox (bovines, man; reservoir hosts unknown) Ectromelia (mice) Monkeypox (monkeys, man; reservoir hosts unknown) Rabbitpox (rabbits) Variola (man)	
Genus	Orf subgroup	*Parapoxvirus*
Type species	Orf virus	–
Main characteristics	One molecule dsDNA, MW = 85×10^6. Virion ovoid, 220–300 nm × 140–170 nm; external coat and filaments are thicker than in vaccinia virions and are arranged as a regular spiral coil consisting of a single thread. Members show serological cross-reactivity; infected cells do not produce hemagglutinin. Viruses of ungulates that may infect man; infectivity is ether-sensitive.	
Other members	Bovine pustular stomatitis Chamois contagious ecthyma Milker's node	
Genus	Fowlpox subgroup	*Avipoxvirus*
Type species	Fowlpox virus	–
Main characteristics	One molecule dsDNA, MW = 200×10^6. Members show serological cross-reactivity. Infectivity is ether-resistant. Type A inclusion bodies contain much lipid, and infected cells do not produce hemagglutinin. Viruses of birds. Mechanical transmission by arthropods is common.	

Taxonomic status	*English vernacular name*	*International name*
Other members	Canary pox Junco pox Pigeon pox Quail pox Sparrow pox Starling pox Turkey pox	
Genus	Sheep pox subgroup	*Capripoxvirus*
Type species	Sheep pox virus	–
Main characteristics	Virions longer and narrower than vaccinia virions; infectivity is ether-sensitive; members show serological cross-reactivity, but produce no hemagglutinin. Viruses of ungulates. Mechanical transmission by arthropods occurs.	
Other members	Goat pox Lumpy skin disease (Neethling)	
Genus	Myxoma subgroup	*Leporipoxvirus*
Type species	Myxoma virus	–
Main characteristics	DNA MW $=150\times10^6$. Serological cross-reactivity occurs between members; infected cells do not produce hemagglutinin. Infectivity is ether-sensitive. Viruses of leporids and squirrels. Mechanical transmission by arthropods is common.	
Other members	Hare fibroma Rabbit (Shope) fibroma Squirrel fibroma	
Genus	Swinepox subgroup	*Suipoxvirus*
Type species	Swinepox virus	–
Main characteristics	Viruses of swine. Limited host range. Infection marked by several types of cytoplasmic inclusion and vacuolation of nuclei.	
Other members	None	
Subfamily	Poxviruses of insects	***Entomopoxvirinae***
Main characteristics	One molecule dsDNA, MW $=140–240\times10^6$. Brick-shaped or ovoid virion 170–250 nm $\times$ 300–400 nm. Virions contain four enzymes found in vertebrate poxviruses. Virions of several morphological types (see below) with globular surface units that give them a mulberry-like appearance; some have only one lateral body, others two. Particles may be occluded in crystalline protein occlusion bodies. Multiplication mainly in cytoplasm of	

Taxonomic status	English vernacular name	International name
	leukocytes and adipose cells. No serological relationships between viruses of the genera or between insect and vertebrate poxviruses. Viruses of insects which probably do not multiply in vertebrates.	
Probable genera (based on morphology of virions, host range, and molecular weight of genome)	Genus A Genus B Genus C	– – –
Probable genus	A	–
Type species	*Melolontha melolontha* (Coleoptera) entomopoxvirus	–
Main characteristics	One molecule dsDNA, MW = $170–240 \times 10^6$. Virions ovoid, 450×250 nm, with one lateral body and unilateral concave core; surface with globular units 22 nm in diameter.	
Other members	Similar viruses recovered from the following Coleoptera: *Anomala cuprea* *Aphodius tasmaniae* *Demodema boranensis* *Dermolepida albohirtum* *Figulus sublaevis* *Geotrupes sylvaticus* *Orthmonius batesi* *Phyllopertha horticola*	
Probable genus	B	–
Type species	*Amsacta moorei* (Lepidoptera) entomopoxvirus	–
Main characteristics	One molecule dsDNA, MW = $132–142 \times 10^6$. G+C content about 26%. Virions ovoid, 350×250 nm, with a sleeve-shaped lateral body and cylindrical core; surface with globular units 40 nm in diameter.	
Other members	Similar viruses recovered from the following Lepidoptera: *Acrobasis zelleri* *Choristoneura biennis* *Choristoneura conflicta* *Choristoneura diversuma* *Chorizagrotis auxiliaris* *Operophtera brumata* *Oreopsyche angustella* and from the following Orthoptera: *Melanoplus sanguinipes*	
Probable genus	C	–

Taxonomic status	English vernacular name	International name
Type species	*Chironomus luridus* (Diptera) entomopoxvirus	–
Main characteristics	One molecule dsDNA, MW = 165–250 × 10^6. Virions brick-shaped, 320 nm × 230 × 110 nm, with two lateral bodies and biconcave core.	
Other members	Similar viruses recovered from the following Diptera: *Aedes aegypti* *Camptochironomus tentans* *Chironomus attenuatus* *Chironomus plumosus* *Goeldichironomus holoprasimus*	
Other members of family ***Poxviridae*** (vertebrate hosts) not yet allocated to genera	Carnivorepox, Elephantpox } related to cowpox Molluscum contagiosum (human) Raccoonpox – probable orthopoxvirus Tanapox, Yaba monkey tumor pox (monkey) } serologically related	
Derivation of names	pox: from plural of pock (Old English *poc, pocc-*), 'pustule, ulcer' ortho: from Greek *orthos*, 'straight, correct' avi: from Latin *avis*, 'bird' capri: from Latin *caper, capri*, 'goat' lepori: from Latin *lepus, leporis*, 'hare' para: from Greek *para*, 'by the side of' entomo: from Greek *entomon*, 'insect' sui: from Latin *sus*, 'swine'	

References

Baxby, D.: Identification and interrelationships of the variola/vaccinia subgroup of poxviruses. Prog. med. Virol., vol. 19, pp. 215–246 (Karger, Basel 1975).

Baxby, D.: Poxvirus hosts and reservoirs. Archs Virol. *55:* 169–179 (1977).

Bergoin, M.; Dales, S.: Comparative observations on poxviruses of invertebrates and vertebrates; in Maramorosch, Kurstak, Comparative virology, pp. 169–205 (Academic Press, New York 1971).

Fenner, F.; Pereira, H. G.; Porterfield, J. S.; Joklik, W. K.; Downie, A. W.: Family and generic names for viruses approved by the International Committee on Taxonomy of Viruses, June 1974. Intervirology *3:* 193–198 (1974).

Granados, R. R.: Insect poxviruses: pathology, morphology and development. Misc. Publs ent. Soc. Am. *9:* 73–94 (1973).

Joklik, W. K.: The poxviruses. A. Rev. Microbiol. *22:* 359–390 (1968).

Mayr, A.; Mahnel, H.; Munz, E.: Systemisierung und Differenzierung der Pockviren. Zentbl. VetMed. *B19:* 69–88 (1972).

Moss, B.: Reproduction of poxviruses; in Fraenkel-Conrat, Wagner, Comprehensive virology, vol. 3, pp. 405–474 (Plenum, New York 1974).

Taxonomic status	*English vernacular name*	*International name*
Family	Herpesvirus group	***HERPESVIRIDAE***

Main characteristics

A. Properties of the Virus Particle

Nucleic acid: One molecule of linear dsDNA. MW = $80–150 \times 10^6$; 35–75% G+C.
Protein: More than 20 structural polypeptides, MWs = 12,000–>220,000.
Lipid: Exact percentage of total weight unknown, probably variable; located in virion envelope.
Carbohydrate: Exact percentage of total weight unknown, identified largely as covalently linked to envelope proteins.
Physicochemical properties: MW $> 1{,}000 \times 10^6$. Buoyant density in CsCl = 1.20–1.29 g/cm^3.
Morphology: The virion, 120–200 nm in diameter, consists of 4 structural components. The core consists of a fibrillar spool on which the DNA is wrapped. The ends of the fibers are anchored to the underside of the capsid shell. The capsid, 100–110 nm in diameter, has 162 capsomers arranged as an icosahedron. It has 5 capsomers on each edge. It contains 150 hexameric and 12 pentameric capsomers. The capsomers are hexagonal in cross-section and contain a hole running half-way down the long axis. The tegument surrounding the capsid consists of globular material which is frequently asymmetrically distributed and may be variable in amount. The envelope, a bilayer membrane surrounding the tegument, has surface projections. The intact envelope is impermeable to negative stain.
Antigenic properties: Neutralizing antibody reacts with major viral glycoproteins located in the viral envelope. An Fc receptor may be present in the envelope.
Effects on cells: Fusion and agglutination occur rarely or under very special conditions in the absence of replication.

B. Replication

Entry: The viral envelope adsorbs to receptors on the plasma membrane of the host cell, ultimately fuses with the membrane, and releases the capsid into the cytoplasm. A DNA-protein complex is then translocated into the nucleus.
Replication: Viral DNA is transcribed in the nucleus. mRNA generated from these transcripts is translated in the cytoplasm. Viral DNA is replicated in the nucleus and is spooled into preformed, immature nucleocapsids.
Maturation and egress: The ability to infect cells is acquired as the capsids acquire the envelope by budding through the inner lamella of the nuclear membrane. The virus accumulates in the space between the inner and outer lamellae of the nuclear membrane and in the cisternae of the endoplasmic reticulum. Virus particles are released by transport to the surface through the modified endoplasmic reticulum.

C. Biological Aspects

Host range: Each virus has its own host range; this host range may vary considerably both in nature and in the laboratory. Herpesviruses occur in both warm- and cold-blooded vertebrates and in invertebrates. Some herpesviruses have been reported to induce neoplasia both in their natural hosts and in experimental animals. In cell culture, herpesviruses have been reported to convert cell strains into continuous cell lines which may cause invasive tumors in appropriate experimental hosts.
Transmission: For many herpesviruses transmission is by contact between moist mucosal surfaces. Some herpesviruses can be transmitted transplacentally, intrapartum, via breast milk, or by blood transfusions, and some are probably also transmitted by airborne and waterborne routes.

Taxonomic status	*English vernacular name*	*International name*
	Association of virus with host: Herpesviruses may remain latent in their primary host fo the lifetime of those hosts; cells harboring latent virus may vary depending on the virus.	
Subfamilies	Herpes simplex virus group Cytomegalovirus group Lymphoproliferative virus group	***Alphaherpesvirinae*** ***Betaherpesvirinae*** ***Gammaherpesvirinae***
Subfamily	Herpes simplex virus group	***Alphaherpesvirinae***
Main characteristics	A. Properties of the Virus Particle Nucleic acid: DNA MW = 85–110 × 10^6. The sequences from both or either terminus ar present in an iverted form internally. The DNA packaged in virions consists of two or fou isometric forms. Natural isolates may differ in the presence and position of restrictio enzyme cleavage sites. B. Replication Relatively short (<24 h) replicative cycle. Latent virus infection frequently demonstrate in ganglia. C. Biological Aspects Host range: Variable, from very wide to very narrow. In vitro also variable. Cytopathology: Rapid spread of infection in cell culture resulting in mass destruction c susceptible cells. Establishment of carrier cultures of susceptible cells harboring nondefec tive genomes difficult to accomplish. Latent infections: Frequently, but not exclusively, in ganglia.	
Genera	Human herpesvirus 1 group Suid herpesvirus 1 group	– –
Genus	Human herpesvirus 1 group	–
Type species	Human (alpha) herpesvirus 1 (herpes simplex virus type 1)	–
Main characteristics	A. Properties of the Virus Particle Nucleic acid: MW = 96 × 10^6; 67% G + C. Sequences from both termini are repeated in a inverted form internally; virion DNA exists in 4 isomeric forms and shares 50% of i sequences with human (alpha) herpesvirus 2 DNA under stringent hybridization condition Proteins: >24 structural proteins, including 5 major glycoproteins. Natural isolates ma differ with respect to electrophoretic mobility of their polypeptides. Antigenic properties: At least 3 glycoproteins are capable of inducing neutralizing antibod B. Biological Aspects Host range: Recovered in nature only from human infection. Experimental host range ve wide.	
Other members	Human (alpha) herpesvirus 2 (herpes simplex virus type 2) Bovid (alpha) herpesvirus 2 (bovine mammillitis)	

Taxonomic status	*English vernacular name*	*International name*
Genus	Suid herpesvirus 1 group	–
Type species	Suid (alpha) herpesvirus 1 (pseudorabies)	–
Main characteristics	A. Properties of the Virus Particle Nucleic acid: MW = 92×10^6; 72% G+C. Sequences from one terminus are repeated in an inverted form internally. Virion DNA exists in 2 isomeric forms. B. Biological Aspects Host range: Recovered from a wide range of species in nature. Experimental host range also very wide.	
Other member	Equid (alpha) herpesvirus 1 (equine abortion)	
Probable members of the ***Alphaherpesvirinae***	Cercopithecid herpesvirus 1 (B) Equid herpesvirus 3 (coital exanthema) Felid herpesvirus 1 (feline herpesvirus) Human (alpha) herpesvirus 3 (varicella-zoster)	
Possible members of the ***Alphaherpesvirinae***	Equid herpesvirus 2 (slow-growing equine herpesvirus) Canid herpesvirus 1 (canine herpesvirus)	
Subfamily	Cytomegalovirus group	***Betaherpesvirinae***
Main characteristics	A. Properties of the Virus Particle Nucleic acid: DNA MW = $130–150 \times 10^6$; 56% G+C. Sequences from either or both termini may be present in an inverted form internally. B. Replication Relatively slow reproductive cycle (>24 h). Slowly progressing lytic foci in cell culture. Enlargement of the infected cell in vivo and often in vitro (cytomegalia). Inclusion bodies containing DNA may be present in the nuclei and cytoplasm late in infection. Carrier cultures easily established. C. Biological Aspects Host range: In vivo – narrow, frequently restricted to the species or genus to which the host belongs. In vitro – replicates best in fibroblasts, although exceptions exist. Latent infections: Possibly in secretory glands, lymphoreticular cells, and kidneys and other tissues.	
Genera	Human cytomegalovirus group Murine cytomegalovirus group	– –
Genus	Human cytomegalovirus group	–
Type species	Human (beta) herpesvirus 5 (human cytomegalovirus)	–

Taxonomic status	*English vernacular name*	*International name*
Main characteristics	DNA MW = 150×10^6. Virus recovered only from human infections. Experimental host range narrow; grows best in human fibroblasts and less well in certain human lymphoblastoid cells.	
Genus	Murine cytomegalovirus group	–
Type species	Murid (beta) herpesvirus 1 (mouse cytomegalovirus)	–
Main characteristics	DNA MW = 132×10^6.	
Possible members of the ***Betaherpesvirinae***	Suid herpesvirus 2 (pig cytomegalovirus) Murid herpesvirus 2 (rat cytomegalovirus) Caviid herpesvirus 1 (guinea pig cytomegalovirus)	
Subfamily	Lymphoproliferative virus group	***Gammaherpesvirinae***
Main characteristics	A. Properties of the Virus Particle Nucleic acid: DNA MW = $85–110 \times 10^6$; both ends of the molecule contain reiterated sequences that are not repeated internally. B. Replication Duration of the replication cycle is variable. All members of this group will replicate in lymphoblastoid cells, and some will also cause lytic infections in some types of epithelioid and fibroblastic cells. Viruses are specific for either B- or T-lymphocytes; in the lymphocyte infection is frequently arrested at a prelytic stage, with persistence and minimum expression of the viral genome in the cell (lymphocytic infection), or at a lytic stage, causing death of the cell without production of complete virions. Latent virus is frequently demonstrated in lymphoid tissue. C. Biological Aspects Host range: Narrow; experimental hosts usually limited to the same order as the host it naturally infects. Cytopathology: Variable.	
Genus	–	Un-named
Type species	Human (gamma) herpesvirus 4 (Epstein-Barr)	–
Main characteristics	DNA MW not greater than 110×10^6; some isolates lack approximately 10×10^6 of DNA situated at a specific site. Virus shows specificity for B-lymphocytes.	
Probable members of the ***Gammaherpesvirinae***	Herpesvirus ateles Herpesvirus saimiri	

Taxonomic status	*English vernacular name*	*International name*
Possible members of the ***Gammaherpesvirinae***	Gallid herpesvirus 1 (Marek's disease herpesvirus) Gallid herpesvirus 2 (turkey herpesvirus) Leporid herpesvirus 1 (rabbit herpesvirus)	

Derivation of name	herpes:	from Greek *herpes, herpetos,* 'creeping, crawling creature'; from nature of herpes febrilis lesions

References

Bornkamm, G.W.; Delius, H.; Fleckenstein, B.; Werner, F.-J.; Mulder, C.: Structure of *Herpesvirus saimiri* genomes: arrangement of heavy and light sequences in the M genome. J. Virol. *19:* 154–161 (1976).

Buchman, T.G.; Roizman, B.: Anatomy of bovine mammillitis DNA. II. Size and arrangements of the deoxynucleotide sequences. J. Virol. *27:* 239–254 (1978).

Fleckenstein, B.; Bornkamm, G.W.; Mulder, C.; Werner, F.-J.; Daniel, M.D.; Falk, L.A.; Delius, H.: *Herpesvirus ateles* DNA and its homology with *Herpesvirus saimiri* nucleic acid. J. Virol. *25:* 361–373 (1978).

Given, D.; Kieff, E.: DNA of Epstein-Barr virus. IV. Linkage map of restriction enzyme fragments of the B95-8 and W91 strains of Epstein-Barr virus. J. Virol. *28:* 524–542 (1978).

Kilpatrick, B.A.; Huang, E.-S.: Human cytomegalovirus genome: partial denaturation map and organization of genome sequences. J. Virol. *24:* 261–276 (1977).

Mosmann, T.R.; Hudson, J.B.: Some properties of the genome of murine cytomegalovirus (MCV). Virology *54:* 135–149 (1973).

Pereira, L.; Cassai, E.; Honess, R.W.; Roizman, B.; Terni, M.; Nahmias, A.: Variability in the structural polypeptides of herpes simplex virus 1 strains: potential application in molecular epidemiology. Infect. Immunity *13:* 211–220 (1976).

Roizman, B.: The structure and isomerization of herpes simplex virus genomes. Cell *16:* 481–494 (1979).

Roizman, B.; Carmichael, L.E.; Deinhardt, F.; de-The, G.; Nahmias, A.J.; Plowright, W.; Rapp, F.; Sheldrick, P.; Takahashi, M.; Wolf, K.: Herpesviridae: definition, provisional nomenclature, and taxonomy. Intervirology *16:* 201–217 (1981).

Taxonomic status	English vernacular name	International name
Family	Baculovirus group	***BACULOVIRIDAE***

Main characteristics

A. Properties of the Virus Particle

Nucleic acid: Single molecule of circular supercoiled dsDNA; MW$\simeq$58–110 × 10^6; 8–15% of particle weight. G+C content is variable from 28 to 59%.

Protein: Virions are structurally complex and contain at least 10–25 polypeptides; MW $\simeq$10–160 × 10^3. Alkaline protease associated with occlusions isolated from infected insects The major structural protein of the occlusion body (where present) consists of a singl polypeptide, viral coded, with MW=25–33 × 10^3. This protein is referred to as polyhedri for nuclear polyhedrosis viruses and granulin for granulosis viruses.

Lipid: Present.

Carbohydrate: Present.

Physicochemical properties: MW not accurately determined. Probably 1,000 × 10^6 fo enveloped single nucleocapsids. Density of nucleocapsid in CsCl=1.47 g/cm^3 and of th virion, 1.18–1.25 g/cm^3. Ether and heat labile.

Morphology: Virions consist of one or more rod-shaped nucleocapsids enclosed within single envelope. Enveloped single nucleocapsids are bacilliform in shape, about 40–60 nm × 200–400 nm.

Antigenic properties: Cross-reacting antigenic determinants exist on the virion structura proteins between and within subgroups. Present on the major subunit of polyhedrin an granulin polypeptides are cross-reacting antigenic determinants.

B. Replication

Viruses of subgroups A and C replicate exclusively in the nucleus. Those of subgroups replicate largely in the nucleus, but replication can occur in the cytoplasm. Virions ma be occluded or non-occluded, depending upon the time or sequence in the infection cycl or the virus species. Virions of subgroups A and B can be occluded in the crystallin protein occlusion body, which may be polyhedral in shape and contain many virus part cles (subgroup A), or may be ovicylindrical and contain only one or rarely two particle (subgroup B). Subgroup C viruses have no occlusion bodies. For all subgroups cell-to cell spread is presumably by means of non-occluded or extracellular virus.

C. Biological Aspects

Baculoviruses have been isolated from Arthropoda: Insecta, Arachnida and Crustace Transmission: (i) natural – horizontal, by contamination of food, etc.; (ii) vertical on th egg; (iii) experimental – by injection of insects or by infection or transfection of cell culture

Taxonomic status	English vernacular name	International name
Genus		*Baculovirus*
Subgroups	A: nuclear polyhedrosis virus	–
	B: granulosis virus	–
	C (proposed): non-occluded rod-shaped nuclear viruses	–
	D (proposed): non-occluded nuclear viruses with a polydisperse DNA genome	–
Subgroup	A: nuclear polyhedrosis virus	–

Taxonomic status	*English vernacular name*	*International name*
Type species	*Autographa californica* nuclear polyhedrosis virus (Ac*M*NPV)	–
Main characteristics	*Autographa californica* nuclear polyhedrosis virus (Ac*M*NPV) is representative of the *M*NPV subtype, where the virions may contain one to many nucleocapsids within a single viral envelope (*M*NPV). All species have many virions per occlusion body.	
Other members	*Bombyx mori* nuclear polyhedrosis virus (Bm*S*NPV) is representative of the *S*NPV subtype where only single nucleocapsids are enveloped with many virions incorporated per occlusion body. Similar viruses isolated from seven insect orders and from Crustacea.	
Subgroup	B: granulosis virus	–
Type species	*Trichoplusia ni* granulosis virus (TnGV)	–
Main characteristics	Enveloped single nucleocapsid with one virion per occlusion body.	
Other members	*Plodia interpunctella* granulosis virus (PiGV), and similar viruses from about 50 species in the Lepidoptera.	
Proposed subgroup	C: non-occluded rod-shaped nuclear viruses	–
Type species	*Oryctes rhinoceros* virus	–
Main characteristics	Enveloped single nucleocapsids. No occlusion bodies produced.	
Proposed subgroup	D: non-occluded nuclear viruses with a polydisperse DNA genome. Viruses with double-stranded circular, polydisperse DNA genome; replication in nuclei of calyx epithelial cells (ovaries of parasitoid hymenoptera); enveloped; transmission invariant, probably vertical in egg.	
Type species and characteristics	Type D1: Type species *Hyposoter exiguae* virus. Genome packaged into quasicylindrical nucleocapsids of uniform dimension with each nucleocapsid surrounded by 2 envelopes. Type D2: Type species *Apanteles melanoscelus* typical of braconid viruses. Genome packaged into cylindrical nucleocapsids of variable length, with single or multiple nucleocapsids surrounded by a single envelope.	
Other possible members of genus *Baculovirus*	A diverse group based upon morphological variation of virus structure which requires further delineation into distinct subgroups as more data become available. These are virus particles with similar general structure to baculoviruses isolated from mites, Crustacea and Coleoptera. Putative baculoviruses have been observed in a fungus *(Strongwellsea magna)*, a spider, the European crab *(Carcinus maenas)*, and the blue crab *(Callinectes sapidus)*.	
Derivation of name	baculo: from Latin *baculum*, 'stick', from morphology of virion	

References

Arnott, H.J.; Smith, K.M.: An ultrastructural study of the development of a granulosis virus in the cells of the moth *Plodia interpunctella* (Hbn.). J. Ultrastruct. Res. *21:* 251–268 (1968).

Carstens, E.B.; Tjia, S.T.; Doerfler, W.: Infection of *Spodoptera frugiperda* cells with *Autographa californica* nuclear polyhedrosis virus. Virology *99:* 386–398 (1979).

Couch, J.A.: An enzootic nuclear polyhedrosis virus of pink shrimp: ultrastructure, prevalence and enhancement. J. invertebr. Pathol. *24:* 311–331 (1974).

Granados, R.R.; Lawler, K.A.: *In vivo* pathway of *Autographa californica* baculovirus invasion and infection. Virology *108:* 297–308 (1981).

Harrap, K.A.: The structure of nuclear polyhedrosis viruses. III. Virus assembly. Virology *50:* 133–139 (1972).

Harrap, K.A.; Payne, C.C.: The structural properties and identification of insect viruses. Adv. Virus Res. *25:* 273–355 (1979).

Harrap, K.A.; Payne, C.C.; Robertson, J.S.: The properties of three baculoviruses from closely related hosts. Virology *79:* 14–31 (1977).

Jewell, J.E.; Miller, L.K.: DNA sequence homology relationship among six lepidopteran nuclear polyhedrosis viruses. J. gen. Virol. *48:* 161–175 (1980).

Jurkovicova, M.; van Touw, J.H.; Sussenbach, J.S.; Ter Schegget, J.: Characterization of the nuclear polyhedrosis virus DNA of *Adoxophyes orana* and of *Barathra brassicae*. Virology *93:* 8–19 (1979).

Krell, P.J.; Stoltz, D.B.: Unusual baculovirus of the parasitoid wasp *Apanteles melanoscelus:* isolation and preliminary characterization. J. Virol. *29:* 1118–1130 (1979).

Krell, P.J.; Stoltz, D.B.: Virus-like particles in the ovary of an ichneumonid wasp: purification and preliminary characterization. Virology *101:* 408–418 (1980).

Maruniak, J.E.; Summers, M.D.: Comparative peptide mapping of baculovirus polyhedrins. J. invertebr. Pathol. *32:* 196–201 (1978).

Payne, C.C.: The isolation and characterization of a virus from *Oryctes rhinoceros*. J. gen. Virol. *25:* 105–116 (1974).

Payne, C.C.; Kalmakoff, J.: Alkaline protease associated with virus particles of a nuclear polyhedrosis virus: assay, purification and properties. J. Virol. *26:* 84–92 (1978).

Smith, G.E.; Summers, M.D.: Analysis of baculovirus genomes with restriction endonucleases. Virology *89:* 517–527 (1978).

Smith, G.E.; Summers, M.D.: Restriction maps of five *Autographa californica M*NPV variants, *Trichoplusia ni M*NPV, and *Galleria mellonella M*NPV DNAs with endonucleases *Sma*I, *Kpn*I, *Bam*HI, *Sac*I, *Xho*I, and *Eco*RI. J. Virol. *30:* 828–838 (1979).

Smith, G.E.; Summers, M.D.: The application of a novel radioimmunoassay to identify baculovirus structural proteins that share interspecies antigenic determinants. J. Virol. *39:* 125–137 (1981).

Summers, M.D.; Hoops, P.: Radioimmunoassay analysis of baculovirus granulins and polyhedrins. Virology *103:* 89–98 (1980).

Summers, M.D.; Smith, G.E.; Krell, J.D.; Burand, J.P.: Physical maps of *Autographa californica* and *Rachiplusia ou* nuclear polyhedrosis virus recombinants. J. Virol. *34:* 693–703 (1980).

Tweeten, K.A.; Bulla, L.A., Jr.; Consigli, R.A.: Characterization of an extremely basic protein derived from granulosis virus nucleocapsids. J. Virol. *33:* 866–876 (1980a).

Tweeten, K.A.; Bulla, L.A., Jr.; Consigli, R.A.: Structural polypeptides of the granulosis virus of *Plodia interpunctella*. J. Virol. *33:* 877–886 (1980b).

Vlak, J.M.; Odink, K.G.: Characterization of *Autographa californica* nuclear polyhedrosis virus deoxyribonucleic acid. J. gen. Virol. *44:* 333–347 (1979).

Vlak, J.M.; Smith, G.E.: Orientation of the genome of *Autographa californica* nuclear polyhedrosis virus: a proposal. J. Virol. *41:* 1118–1121 (1982).

Wood, H.A.: *Autographa californica* nuclear polyhedrosis virus-induced proteins in tissue culture. Virology *102* 21–27 (1980).

Taxonomic status	*English vernacular name*	*International name*
Family	Mycoplasma virus type 2 phages	***PLASMAVIRIDAE***
Genus	Mycoplasma virus type 2 phages	*Plasmavirus*
Type species	Phage MV-L2	–
Main characteristics	A. Properties of the Virus Particle Nucleic acid: One piece of circular supercoiled dsDNA; MW$\simeq 7.6 \times 10^6$. Protein: At least 7 proteins; MW$\simeq$19,000–68,000. Lipid: Present in envelope, 0.08–0.10 μmol lipid phosphorus/mg protein; lipids and fatty acids similar to those in host cell membranes. Morphology: Rounded, slightly pleomorphic, flexible envelope, about 80 (range 50–120) nm in diameter; size range is due to three distinct populations of particles produced during infection. Sections show a small, densely stained core inside the envelope. Physicochemical properties: Infectivity is ether-, chloroform-, detergent-, and heat-sensitive. B. Replication Mature viruses released by budding; no lysis, host survives as lysogen. C. Biological Aspects Host range: *Acholeplasma*	
Other member	1307	
Possible members	MV-Lg-pS2L, v1, v2, v4, v5, v7	
Derivation of name	plasma: from Greek *plasma,* 'shaped product'	

References

Ackermann, H.-W.: Cubic, filamentous, and pleomorphic bacteriophages; in Laskin, Lechevalier, CRC handbook of microbiology; 2nd ed., vol. II, pp. 673–682 (CRC Press, West Palm Beach 1978).

Maniloff, J.; Cadden, S.P.; Putzrath, R.M.: Maturation of an enveloped budding phage: mycoplasmavirus L2; in Dubow, Bacteriophage assembly, pp. 503–513 (Liss, New York 1981).

Maniloff, J.; Das, J.; Christensen, J.R.: Viruses of mycoplasmas and spiroplasmas. Adv. Virus Res. *21:* 343–380 (1977).

Taxonomic status	English vernacular name	International name
Family	Icosahedral cytoplasmic deoxyriboviruses	***IRIDOVIRIDAE***

Main characteristics

A. Properties of the Virus Particle

Nucleic acid: One molecule of linear dsDNA, MW$\simeq$100–250 $\times$ 10^6. Possibly two molecules in some viruses. 12–30% by weight of the virus particle.

Protein: 13–25 structural polypeptides by one-dimensional PAGE, with MWs ranging from 10–250 $\times$ 10^3. Most members that have been examined possess several virion-associated enzymes, in particular an active protein kinase.

Lipid: Unenveloped particles contain 5–9% lipid (predominantly phospholipid) as an integral part of the icosahedral shell. Some members have an additional plasma-membrane-derived envelope.

Carbohydrate: None has been reported.

Physicochemical properties: MW of virions = 500–2,000 $\times$ 10^6; S_{20w} = 1,300–4,450; density = 1.16–1.35 g/cm^3; members of *Iridovirus* and *Chloriridovirus* genera resistant to ether, all others sensitive to ether and nonionic detergents; stable at pH 3–10 and at 4° for several years; inactivated by 15–30 min at 55°.

Morphology: Icosahedral, 125–300 nm diameter; spherical nucleoprotein core surrounded by membrane consisting of lipid modified by morphological protein subunits; the released virions of some genera possess a plasma-membrane-derived envelope that is not required for infectivity.

Antigenic properties: Antibodies prepared against virions appear to be non-neutralizing, but are useful in establishing relationships between species.

Effects on cells: Generally cytocidal; most members rapidly inhibit host cell macromolecular synthesis.

B. Replication

Virus entry is by pinocytosis, with uncoating in phagocytic vacuoles. The host cell nucleus appears to be required for transcription and replication of DNA, but some DNA synthesis, and the assembly of virions into mature particles, takes place in the cytoplasm, where paracrystalline inclusion bodies are readily observed. Release is by lysis or budding. Virions that bud from the host acquire a plasma- or endoplasmic reticulum-derived envelope, but most virus remains cell-associated and unenveloped virions are infectious.

C. Biological Aspects

Host range: Many species appear to have a restricted host range in vivo and in vitro; exceptions are the genus *Ranavirus* (frog virus 3), which grows in a wide variety of cultured cells, and *Iridovirus* (*Tipula* iridescent virus), which infects a wide range of insects.

Transmission: Both horizontal and vertical; African swine fever virus can multiply and be transmitted transovarially in ticks.

Genera	Small iridescent insect virus group	*Iridovirus*
	Large iridescent insect virus group	*Chloriridovirus*
	Frog virus group	*Ranavirus*
	African swine fever virus genus	–
	Lymphocystis disease virus group	–

Genus	Small iridescent insect virus group	*Iridovirus*

Taxonomic status	*English vernacular name*	*International name*
Proposed type species	*Tipula* iridescent virus	–
Main characteristics	Particles 120 nm diameter. Complex icosahedral shell contains lipid, but integrity is protected under protein capsid as infectivity is not sensitive to ether. Infected larvae and purified virus pellets produce a blue iridescence.	
Other members	Insect iridescent viruses 1, 2, 6, 9, 10, 16–29	
Probable member	*Chironomus plumosus* iridescent	
Possible member	*Octopus vulgaris* disease	
Genus	Large iridescent insect virus group	*Chloriridovirus*
Type species	Mosquito iridescent virus (iridescent virus – type 3, regular strain)	
Main characteristics	Particle diameter = 180 nm. Virus glows with a yellow-green iridescence in infected larvae and in virus pellets.	
Other members	Insect iridescent viruses 3–5, 7, 8, 11–15	
Genus	Frog virus group	*Ranavirus*
Type species	Frog virus 3 (FV3)	–
Main characteristics	Does not cause disease in natural host, adult *Rana pipiens,* but is lethal for tadpoles and Fowler toads; grows in piscine, avian, and mammalian cells from 12 to 32°; structural viral protein causes rapid inhibition of host macromolecular synthesis without interfering with viral replication. DNA-dependent RNA polymerase not found associated with virus particle. DNA contains a high proportion of 5-methyl cytosine and is circularly permuted.	
Other members	Frog viruses 1, 2, 5–24, L2, L4 and L5 Tadpole edema virus from *R. catesbriana* LT 1–4 and T6–T20 from newts T21 from *Xenopus*	
Genus	African swine fever virus group	–
Type species	African swine fever virus	–
Main characteristics	Infects various species of swine (causes severe systemic febrile disease in natural host); grows readily in swine mononuclear cells in vitro and, after adaptation, in several homeothermic cell lines. Virus particles possess hemadsorption ability. The DNA has cross-linked ends. A DNA-dependent RNA polymerase is associated with the virus particle. Requires the host nucleus for replication.	
Genus	Lymphocystis disease virus group	–

Taxonomic status	*English vernacular name*	*International name*
Proposed type species	Lymphocystis disease virus	–

Derivation of names	irido:	from Greek *Iris, Iridos,* goddess whose sign was the rainbow, hence iridescent; 'shining like a rainbow', from appearance of infected larval insects and centrifuged pellets of virions
	chloro:	from Greek *chloros,* 'green'
	rana:	from Latin *rana,* 'frog'

References

Bellett, A.J.D.: The iridescent virus group. Adv. Virus Res. *13:* 225–246 (1968).

Black, D.N.; Brown, F.: Purification and physicochemical characteristics of African swine fever virus. J. gen. Virol. *32:* 509–518 (1976).

Christmas, J.Y.; Howse, H.D.: The occurrence of lymphocystis in *Micropogon undulatus* and *Cynoscian arenius* from Mississippi estuaries. Gulf Res. Rep. *3:* 131–154 (1970).

Elliott, R.M.; Lescott, T.; Kelly, D.C.: Serological relationships of an iridescent virus (type 25) recently isolated from *Tipula* sp. with two other iridescent viruses (types 2 and 22). Virology *81:* 309–316 (1977).

Goorha, R.; Granoff, A.: Icosahedral cytoplasmic deoxyriboviruses; in Fraenkel-Conrat, Wagner, Comprehensive virology, vol. 14, pp. 347–399 (Plenum, New York 1979).

Goorha, R.; Murti, K.G.: The genome of an animal DNA virus (frog virus 3) is circularly permuted and terminally redundant. Proc. natn. Acad. Sci. USA (in press, 1981).

Hess, W.R.: African swine fever virus. Virology Monographs, vol. 9, pp. 1–33 (Springer, New York 1971).

Kelly, D.C.; Robertson, J.S.: Icosahedral cytoplasmic deoxyriboviruses. J. gen. Virol. *20:* suppl., pp. 17–41(1973).

Kelly, D.C.; Tinsley, T.W.: Iridescent virus replication: patterns of nucleic acid synthesis in insect cells infected with iridescent virus types 2 and 6. J. invertebr. Pathol. *24:* 169–178 (1974).

Kuznar, J.; Salas, M.L.; Vinuela, E.: DNA-dependent RNA polymerase in African swine fever virus. Virology *101:* 169–175 (1980).

Ortin, J.; Enjuanes, L.; Vinuela, E.: Cross-links in African swine fever virus DNA. J. Virol. *31:* 579–583 (1979).

Siegel, M.M.; Lopez, D.M.; Beasley, A.R.; Caliguri, L.A.: Virus-cell interaction in lymphocystis disease of fish; in Sanders, Schaeffer, Viruses affecting man and animals, pp. 124–135 (Green, St. Louis 1971).

Stoltz, D.B.: The structure of icosahedral cytoplasmic deoxyriboviruses. II. An alternative model. J. Ultrastruct. Res. *43:* 58–74 (1973).

Wagner, G.W.; Paschke, J.D.; Campbell, W.R.; Webb, S.R.: Proteins of two strains of mosquito iridescent virus. Intervirology *3:* 97–105 (1974).

Willis, D.B.; Granoff, A.: Frog virus 3 DNA is heavily methylated at CpG sequences. Virology *107:* 250–257 (1980).

Taxonomic status	*English vernacular name*	*International name*
Family	Adenovirus family	***ADENOVIRIDAE***

Main characteristics

A. Properties of the Virus Particle

Nucleic acid: Genome is a single linear molecule of dsDNA of MW 20–25 × 10^6 for mammalian strains (M) or 28–30 × 10^6 for avian strains (A). G+C content 48–61% (M) and 54–55% (A). Inverted terminal repetition of ≃100 base pairs allows the formation of single-stranded circles.

Protein: At least 10 polypeptides in virion, MWs ranging from 5 × 10^3 to 120 × 10^3 (M).

Lipid: None

Carbohydrate: Fibers are glycoproteins.

Physicochemical properties: MW≃170 × 10^6. Density in CsCl 1.33–1.34 (M) or 1.32–1.35 g/cm^3 (A). Stable on storage; no inactivation by lipid solvents.

Morphology: Virion is a nonenveloped isometric particle with icosahedral symmetry, 70–90 nm in diameter, with 252 capsomers, 8–9 nm in diameter. 12 vertex capsomers (or penton bases) carry one (M) or two (A) filamentous projections (or fibers) of different length; 240 nonvertex capsomers (or hexons) are different from penton bases and fibers.

Antigenic properties: Antigens at the surface of virion are mainly species-specific: hexon for neutralization, fiber for hemagglutination and hemagglutination-inhibition. Soluble antigens are surplus products of capsid proteins; free hexon mainly reacts as a genus-specific antigen, which is shared by most M strains and differs from the corresponding antigen in A strains. Hexons and other soluble antigens carry genus-, subgroup-, and/or species-specific determinants.

Effects on cells: Characteristic CPE during multiplication in cell cultures. Many species hemagglutinate blood cells of various host species. Many species are oncogenic in experimental animals or may transform cells.

B. Replication

Molecular biology: Two major stages of genome transcription: 'early' (before initiation of viral DNA synthesis when mostly nonvirion polypeptides are synthesized) and 'late' (when viral DNA and structural polypeptides are synthesized). Five separate early gene regions, located on both complementary DNA strands. DNA replication by strand displacement mechanism. Transcription in nucleus followed by processing into mRNAs migrating into cytoplasm. Synthesis of structural polypeptides in the cytoplasm. Assembly of virus particles in nucleus. Slow release after rupture of the damaged cell. Virus infection shuts off host cell DNA, RNA and protein synthesis.

Virus inclusion bodies: Intranuclear inclusions, containing DNA and viral antigen, also virions in paracrystalline array or otherwise.

C. Biological Aspects

Host range: Natural host range mostly confined to one host or closely related animal species; this holds also for cell cultures. Some human adenoviruses cause productive infection in rodent cells with low efficiency. Several species cause tumors in newborn hosts of heterologous species. Subclinical infections are frequent in various virus/host systems.

Transmission: Direct or indirect transmission from throat, feces, eye, or urine.

D. Species

Note: Under new rule 18 (p. 24), the definition and naming of the species listed below is provisional. Final approval can be given by ICTV in 1984.

Definition of species: A species (formerly type) is defined on the basis of its immunological

Taxonomic status	*English vernacular name*	*International name*
	distinctiveness, as determined by quantitative *neutralization* with animal antisera. A species has either no cross-reaction with others or shows a homologous-to-heterologous titer ratio of >16 in both directions. If neutralization shows a certain degree of cross-reaction between two viruses in either or both directions (homologous-to-heterologous titer ratio of 8 or 16), distinctiveness of species is assumed if: (i) the hemagglutinins are unrelated, as shown by lack of cross-reaction on hemagglutination-inhibition; or (ii) substantial biophysical/biochemical differences of the DNAs exist. Subgenera: 34 human adenovirus species are classified according to their structural, biochemical, biological and immunological characteristics into 5 subgenera (formerly subgroups) A to E [*Green* et al., 1979; *Wadell* et al., 1980]. Naming of species: Human adenoviruses are designated by the letter 'h' plus a number, viruses from animals by a 3-letter code from the genus of the respective host plus a number as in the following table.	

Table III. Species in the family ***Adenoviridae***

Hosts		Virus genus	Provisionally approved virus species names
English name	zoological name		
Man	*Homo sapiens*	*Mastadenovirus*	h 1–h 34
Cattle	*Bos taurus*		bos 1– bos 9
Pig	*Sus domesticus*		sus 1 – sus 4
Sheep	*Ovis aries*		ovi 1 – ovi 5
Horse	*Equus caballus*		equ 1
Dog	*Canis familiaris*		can 1
Goat	*Capra hircus*		cap 1
Mouse	*Mus musculus*		mus 1
Fowl	*Gallus domesticus*	*Aviadenovirus*	gal 1 – gal 9
Turkey	*Meleagris gallopavo*		mel 1 – mel 2
Goose	*Anser domesticus*		ans 1
Pheasant	*Phasianus colchicus*		pha 1
Duck	*Anas domestica*		ana 1

Genera	Mammalian adenoviruses Avian adenoviruses	*Mastadenovirus* *Aviadenovirus*
Genus	Mammalian adenoviruses	*Mastadenovirus*
Type species	Human adenovirus 2	h 2
Genus	Avian adenoviruses	*Aviadenovirus*
Type species	Fowl adenovirus 1 (CELO)	gal 1

Taxonomic status	*English vernacular name*		*International name*
Derivation of names	adeno:	from Greek *aden, adenos,* 'gland'; adenoviruses were first isolated from human adenoid tissue	
	avi:	from Latin *avis,* 'bird'	
	mast:	from Greek *mastos,* 'breast' – a by-form is Greek and Latin *mamma,* hence mammalian	

References

Flint, S.J.; Broker, T.R.: Lytic infection by adenoviruses; in Tooze, DNA tumor viruses: molecular biology of tumor viruses; 2nd ed., part 2, pp.443–546 (Cold Spring Harbor Laboratory, Cold Spring Harbor 1980).

Ginsberg, H.S.: Adenovirus structural proteins; in Fraenkel-Conrat, Wagner, Comprehensive virology, vol.13, pp.409–457 (Plenum, New York 1979).

Ginsberg, H.S.; Young, C.S.H.: Genetics of adenoviruses; in Fraenkel-Conrat, Wagner, Comprehensive virology, vol.9, pp.27–88 (Plenum, New York 1977).

Green, M.; Mackey, J.K.; Wold, W.S.M.; Rigden, P.: Thirty-one human adenovirus serotypes (Ad1–31) form five groups (A–E) based upon DNA genome homologies. Virology *93:* 481–492 (1979).

Norrby, E.: Biological significance of structural adenovirus components. Curr. Top. Microbiol. Immunol. *43:* 1–43 (1968).

Norrby, E.; Bartha, A.; Boulanger, P.; Dreizin, R.S.; Ginsberg, H.S.; Kalter, S.S.; Kawamura, H.; Rowe, W.P.; Russell, W.C.; Schlesinger, R.W.; Wigand, R.: Adenoviridae. Intervirology *7:* 117–125 (1976).

Philipson, L.: Adenovirus proteins and their messenger RNAs. Adv. Virus Res. *25:* 357–405 (1979).

Philipson, L.; Pettersson, U.; Lindberg, U.: Molecular biology of adenoviruses. Virology Monographs, vol.14 (Springer, Vienna 1975).

Wadell, G.; Hammarskjöld, M.-L.; Winberg, G.; Varsanyi, T.W.; Sundell, G.: Genetic variability of adenoviruses. Ann. N.Y. Acad. Sci. *354:* 16–42 (1980).

Taxonomic status	English vernacular name	International name
Family	Papovavirus group	***PAPOVAVIRIDAE***
Main characteristics	A. Properties of the Virus Particle Nucleic acid: One molecule dsDNA, circular; $MW = 3–5 \times 10^6$; G+C content 41–49%; 10–12% of virion by weight. Protein: 5–7 polypeptides, $MWs \simeq 10–75 \times 10^3$; low MW components are cellular histones. Lipid: None. Carbohydrate: None. Physicochemical properties: $MW \simeq 25–47 \times 10^6$. $S_{20w} = 240–300$. Buoyant density in $CsCl = 1.32\ g/cm^3$. Resistant to ether, acid, and heat treatment. Morphology: Nonenveloped, icosahedral particles 45–55 nm diameter; 72 capsomers in skew arrangement; filamentous forms occur. Antigenic properties: Different species antigenically distinct by neutralization and HI tests; antisera prepared against disrupted virions detect common antigens shared by the other species belonging to the same genus. Effects on cells: Cytolytic in cells of host of origin; may transform cells from other species; several species of virus hemagglutinate by reacting with neuraminidase-sensitive receptors; no tissue culture systems for papillomaviruses. B. Replication Virions attach to cellular receptors, are engulfed and transported to nucleus; host cell enzymes are derepressed and cellular DNA synthesis is stimulated; expression of viral genome divided into early and late events; host cell histones are incorporated into virions during maturation in nucleus; virions released by lysis of infected cells. C. Biological Aspects Host range: Each virus has its own host range, in nature and in laboratory. Transformation tends to occur in cells which do not support replication of virus. Transmission: Contact and airborne infection. Papillomaviruses may be mechanically transmitted by arthropods.	
Genera	– –	*Papillomavirus* *Polyomavirus*
Genus	–	*Papillomavirus*
Type species	Rabbit (Shope) papilloma virus	–
Main characteristics	MW of $DNA = 5 \times 10^6$; diameter of capsid = 55 nm; G+C ratio 41–45%; $S_{20w} = 300$. Causes papillomas in natural hosts. Each virus species contains a distinct surface antigen, but all members of the genus share a common antigen revealed by disrupting the virions. No tissue culture systems available. May be mechanically transmitted by arthropods.	
Other members	Members of this genus are known for man (9 types), cow (5 types), deer, dog, goat, horse, rat, and sheep.	
Genus	–	*Polyomavirus*

Taxonomic status	*English vernacular name*		*International name*
Type species	Polyoma virus (mouse)		–
Main characteristics	MW of DNA = 3×10^6; diameter of capsid = 45 nm; G+C ratio 41–48%, S_{20w} = 240. Several species hemagglutinate. Whole viruses show no serological cross-reactivity between most species, but a common genus antigen can be detected in disrupted virions of all species. T antigens induced by primate viruses cross-react. Inapparent infections in most hosts. Oncogenic in hosts (chiefly immunodeficient newborn hamsters) which are often different from species of origin of virus. Viral DNA integrates into cellular chromosomes of transformed cells.		
Other members and possible members	*Note:* Although the BK and JC viruses are included in the *Polyomavirus* genus, there is no evidence that these viruses are associated with human neoplasms.		
	BK and JC	human	
	K	mouse	
	RKV	rabbit	
	HaPV	hamster	
	SV40	rhesus monkey	
	STMV	stump-tailed macaque	
	Lymphotropic, LPV	green monkey, human (?)	
	SA12	baboon	
Derivation of names	papova:	sigla from *pa*pilloma, *po*lyoma, and *va*cuolating agent (early name for SV40)	
	papilloma:	from Latin *papilla,* 'nipple, pustule', and Greek suffix *-oma,* used to form nouns denoting 'tumors'	
	polyoma:	from Greek *poly,* 'many', and *-oma,* denoting 'tumors'	

Reference

Melnick, J. L.; Allison, A. C.; Butel, J. S.; Eckhart, W.; Eddy, B. E.; Kit, S.; Levine, A. J.; Miles, J. A. R.; Pagano, J. S.; Sachs, L.; Vonka, V.: Papovaviridae. Intervirology *3:* 106–120 (1974).

Taxonomic status	*English vernacular name*	*International name*
Group	Cauliflower mosaic virus group	***CAULIMOVIRUS***
Type member	Cauliflower mosaic virus (CaMV) (24, 243) (cabbage B, Davis isolate)	–
Main characteristics	A. Properties of the Virus Particle Nucleic acid: One molecule of dsDNA; open circular molecule with single-strand discontinuities (gaps) at specific sites, the transcribed (α) strand with one and the non-transcribed (β) strand with two gaps; two DNAs (isolate Cabb B-S with 8,024 base pairs, and isolate CM1841 with 8,031 base pairs) have been sequenced. Protein: Single coat polypeptide, phosphorylated, $MW \simeq 42 \times 10^3$; degrades (probably during preparation) to give several polypeptides with major component $MW \simeq 37 \times 10^3$. Lipid: None. Carbohydrate: Protein has some glycosylation. Physicochemical properties: $MW \simeq 22.8 \times 10^6$; $S_{20w} \simeq 208$; $D \simeq 0.75 \times 10^{-7}$ cm^2/s; apparent partial specific volume $\simeq 0.704$; density 1.37 g/cm^3 in CsCl; particles very stable. Morphology: Isometric particles $\simeq 50$ nm in diameter. Antigenic properties: Efficient immunogens; serological relationships among some members. B. Replication The site(s) of viral DNA replication and transcription is uncertain but it is possibly in the nucleus. DNA is transcribed asymmetrically from the α strand into at least eight transcripts, the largest being a full-length transcript of the α strand. Virus particles accumulate in large electron-dense inclusion bodies in the cytoplasm. These consist of virus-coded protein ($MW \simeq 60 \times 10^3$) and are characteristic for the group. C. Biological Aspects Host range: Narrow. Transmission: Transmissible experimentally by mechanical inoculation; transmitted by aphids in both a nonpersistent and a semipersistent manner, i.e., bimodally.	
Other members	Carnation etched ring (182) Dahlia mosaic (51) Figwort mosaic *Mirabilis* mosaic Strawberry vein banding (219)	
Possible members	Cassava vein mosaic *Petunia* vein clearing *Plantago* virus (4) Thistle mottle	
Derivation of name	caulimo: sigla from *cauli*flower *mo*saic	

References

Al Ani, R.; Pfeiffer, P.; Lebeurier, G.: The structure of cauliflower mosaic virus. II. Identity and location of the viral polypeptides. Virology *93:* 188–197 (1979).

Frank, A.; Guilley, H.; Jonard, G.; Richards, K.; Hirth, L.: Nucleotide sequence of cauliflower mosaic virus DNA. Cell *21:* 285–294 (1980).

Gardner, R.C.; Howarth, A.J.; Hahn, P.; Brown-Luedi, M.; Shepherd, R.J.; Messing, J.: The complete nucleotide sequence of an infectious clone of cauliflower mosaic virus by M13mp7 shotgun sequencing. Nucl. Acids Res. *9:* 2871–2888 (1981).

Guilfoyle, T.J.: Transcription of the cauliflower mosaic virus genome in isolated nuclei from turnip leaves. Virology *107:* 71–80 (1980).

Shepherd, R.J.; Lawson, R.H.: Caulimoviruses; in Kurstak, Handbook of plant virus infections and comparative diagnosis, pp. 847–878 (Elsevier/North Holland, Amsterdam 1981).

Shepherd, R.J.; Richins, R.; Shalla, T.A.: Isolation and properties of the inclusion bodies of cauliflower mosaic virus. Virology *102:* 389–400 (1980).

Taxonomic status	*English vernacular name*	*International name*
Family	PRD1 phage group	***TECTIVIRIDAE***
Genus	PRD1 phage group	*Tectivirus*
Type species	Phage PRD1	–
Main characteristics	A. Properties of the Virus Particle Nucleic acid: One piece of linear dsDNA; MW$\simeq 9 \times 10^6$, about 14% of particle. Protein: 16–18 proteins for PRD1. Lipid: 10–20% by weight of particle; seems to be located in the inner coat and differ from that of the host; over 60% is phospholipid. Carbohydrate: Not known. Physicochemical properties: Particle weight$\simeq 70 \times 10^6$ (ϕNS11), $S_{20w} \simeq 390$; buoyant den sity in CsCl$\simeq$1.28 g/cm^3. Infectivity is ether- and chloroform-sensitive. Morphology: Icosahedral, about 65 nm diameter. Some show single, 20 nm long spike on vertices. Double capsid consisting of a rigid outer shell 3 nm thick and a flexible inne coat 5–6 nm thick. The latter is destroyed by lipid solvents. Upon nucleic acid ejection a tail-like structure of about 60 nm in length appears. No envelope. B. Replication Virions active on gram-negative bacteria adsorb to tips of plasmid-dependent pili. Lysis C. Biological Aspects Host range: Gram-negative bacteria harboring certain drug-resistance plasmids (entero bacteria, *Acinetobacter, Pseudomonas, Vibrio*) and *Bacillus*.	
Other members	L17, PR3, PR4, PR5, PR772 (gram-negatives), AP50, Bam35, ϕNS11 *(Bacillus)*	
Derivation of name	tecti: from Latin *tectus,* 'covered'.	

References

Ackermann, H.-W.; Roy, R.; Martin, M.; Murthy, M.R.V.; Smirnoff, W.A.: Partial characterization of a cubic *Bacillus* phage. Can. J. Microbiol. *24:* 986–993 (1978).

Bamford, D.H.; Rouhiainen, L.; Takkinen, K.; Söderlund, H.: Comparison of the lipid-containing bacterio phages PRd1, PR3, PR4, PR5, and L17. J. gen. Microbiol. (in press).

Taxonomic status	*English vernacular name*	*International name*
Family	PM2 phage group	***CORTICOVIRIDAE***
Genus	PM2 phage group	*Corticovirus*
Type species	Phage PM2	–
Main characteristics	A. Properties of the Virus Particle Nucleic acid: One piece of circular supercoiled dsDNA; MW$\simeq 6 \times 10^6$; 14% by weight of particle; 43% G+C content. Protein: Four proteins with MWs$=4.6$–43×10^3. Protein I forms spikes; II forms outer shell; inner shell of virion contains a transcriptase (protein IV?). Proteins III and IV behave as proteolipids. Lipid: 12–14% of particles; forms a bilayer between outer and inner shell and differs from that of the host; over 90% is phospholipid. Carbohydrate: None? Physicochemical properties: MW$\simeq 50 \times 10^6$; $S_{20w}=230$; buoyant density in CsCl$=1.28$ g/cm^3. Infectivity is ether-, chloroform-, and detergent-sensitive. Morphology: Icosahedral, about 60 nm diameter, with brush-like spikes on vertices. Multilayered capsid. No envelope. B. Replication Adsorption to cell wall; maturation at cell periphery; no inclusion bodies; lysis. C. Biological Aspects Host range: A marine pseudomonad.	
Possible member	06N 58P *(Vibrio)*	
Derivation of name	cortico: from Latin *cortex, corticis,* 'bark, crust'	

References

Franklin, R.M.; Marcoli, R.; Satake, H.; Schäfer, R.; Schneider, D.: Recent studies on the structure of bacteriophage PM2. Med. Microbiol. Immunol. *164:* 87–95 (1977).

Mindich, L.: Bacteriophages that contain lipid; in Fraenkel-Conrat, Wagner, Comprehensive virology, vol. 12, pp. 271–335 (Plenum, New York 1978).

Taxonomic status	*English vernacular name*	*International name*
–	Tailed phages	–
Main characteristics	Extremely variable in dimensions and physicochemical properties; over 1,800 published descriptions. A. Properties of the Virus Particle Nucleic acid: One piece of linear dsDNA; MW = $12–490 \times 10^6$; 30–63% by weight of particle. G + C content is 28–72% and usually close to that of the host. DNA may contain unusual bases, which replace normal bases partially or completely, and unusual sugars. It may be circularly permuted, terminally redundant, or nicked and may have cohesive ends or strands of different weight. Protein: Virions usually contain many different polypeptides (5–23 ?) (MW $4–155 \times 10^3$). Some proteins are associated with the DNA. Lysozyme is located at tail tip; other enzymes may be present. Lipid: Up to 14% by weight has been reported, but the presence of lipids is controversial. Carbohydrate: Glycoproteins have been reported. Physicochemical properties: MWs = $18–470 \times 10^6$; S_{20w} = 226–1,230; buoyant density in CsCl = 1.41–1.55 g/cm^3. Infectivity is generally ether- and chloroform-resistant. Morphology: Virions consist of head (capsid), tail, and fixation organelles. No envelope. Heads are isometric or elongated. Insofar as known, they are icosahedra or derivatives thereof (proposed triangulation numbers T = 7, T = 9, T = 13, T = 21). Capsomers are seldom visible, and heads usually appear smooth. Isometric heads are 40–180 nm in diameter. Tails are helical and contractile, long and noncontractile, or short. They may have base plates, spikes, or fibers, and undergo functional changes. Some phages have collars, and head or collar appendages. Aberrant structures are frequent. Antigenic properties: Virions are antigenically complex. B. Replication Tailed phages are virulent or temperate. Temperate phages have a vegetative and a prophage state. Prophages are integrated in and replicate synchronously with the host genome, or are in the cytoplasm and behave as plasmids. Transduction and conversion ability. Virions adsorb tail first to cell wall, capsule, flagella, or pili. The cell wall is digested by phage lysozyme. Infecting DNA replicates in a semiconservative way. Replicative intermediates are concatemers of circles. Assembly is complex and includes prohead formation and several pathways for separate phage components. DNA is cut to size and packed into capsids. Maturing phages are usually dispersed through the cell; some form regular arrays. Lysis. C. Biological Aspects Host range: Over 90 genera of bacteria and cyanobacteria.	
Families	Phages with contractile tails Phages with long, noncontractile tails Phages with short tails	***Myoviridae*** Styloviridae (proposed name) ***Podoviridae***

Family	Phages with contractile tails	***MYOVIRIDAE***
Main characteristics	Tail long (80–455 nm) and complex, consisting of a central tube and a contractile sheath separated from the head by a neck. Contraction seems to require ATP. Relatively large capsids.	

Taxonomic status	*English vernacular name*	*International name*
Genus	T-even phage group	–
Type species	Coliphage T2	–
Main characteristics (type species only)	A. Properties of the Virus Particle Nucleic acid: MW$\simeq 120 \times 10^6$; 49% by weight of particle; contains hydroxymethylcytosine instead of thymine; G+C content 36%; contains glucose. DNA is circularly permuted and terminally redundant. Protein: At least 21 polypeptides with MWs = $8\text{–}155 \times 10^3$; 1,600–2,000 copies of major capsid protein (MW$\simeq 43 \times 10^3$); 2 or 3 proteins are located inside the head. Various enzymes are present, e.g., dehydrofolate reductase, thymidylate synthetase. Other constituents: Particles contain ATP, folate and polyamines. Physicochemical properties: MW$\simeq 210 \times 10^6$; S_{20w} = 1,040; buoyant density in CsCl = 1.49 g/cm^3. Infectivity is chloroform-resistant. Morphology: Virions have an elongated head of about 110 × 80 nm and a tail of 113 × 16 nm. Tail has a collar, a base plate, 6 spikes and 6 long fibers. B. Replication Adsorption site is cell wall; virulent infection. Host chromosome breaks down and viral DNA replicates as concatemer. Heads, tails, and tail fibers are assembled in 3 different pathways. C. Biological Aspects Host range: Enterobacteria.	
Other members	T6, C16, DdVI, PST, SMB, SMP2, α1, 3, 3T+, 9/0, 11F, 50, 66F, 5845, 8893 and about 40 others (infecting enterobacteria).	

Other members of the family include the following viruses which will probably be assigned to several genera:

(a) Heads isometric: K19, O1, P1, P2, ViI, 121 (enterobacteria), PIIBNV6 *(Agrobacterium)*, A6 *(Alcaligenes)*, G, PBS1, SP3, SP8, SP–15, SP50 *(Bacillus)*, HM3 *(Clostridium)*, I3 *(Mycobacterium)*, CP1, PB–1, PP8, φKZ, φW–14, 12S *(Pseudomonas)*, CT4, e, m, WT1 *(Rhizobium)*, Twort *(Staphylococcus)*, RZh *(Streptococcus)*, and XP5 *(Xanthomonas)*.

(b) Heads elongated: 9266, 16-19 (enterobacteria), 108/106 *(Thermomonospora)*.

Family	Phages with long, noncontractile tails	Proposed name: Styloviridae
Main characteristics	Tail long (64 ?–539 nm) and noncontractile.	

Genus	λ phage group	–
Type species	Coliphage λ	–
Main characteristics (type species only)	A. Properties of the Virus Particle Nucleic acid: MW$\simeq 33 \times 10^6$; 54% by weight of particle; contains 51% G+C; has cohesive ends.	

Taxonomic status	*English vernacular name*	*International name*
	Protein: Nine structural proteins; MWs = $17–130 \times 10^3$; about 420 copies of major capsid protein (MW = 38×10^3). Physicochemical properties: MW $\simeq 60 \times 10^6$; $S_{20w} = 388$; buoyant density in CsCl = 1.49 g/cm^3. B. Replication Adsorption site is cell wall. Temperate infection. Infecting DNA circularizes and replicates or integrates into the host genome. Replicating DNA forms concatemers. No breakdown of host DNA. Heads and tails assemble in 2 pathways. C. Biological Aspects Host range: Enterobacteria.	
Other members	PA2, φD328, φ80	
Possible member	T1	

Other members of family include the following viruses which will probably be assigned to several genera:

(a) Heads isometric: Jersey, T5, ViII, β4, χ (enterobacteria), PS8 *(Agrobacterium)*, A5/A6, 8764 *(Alcaligenes)*, PBP1, SPP1, SPβ, α, φ105, II *(Bacillus)*, HM7 *(Clostridium)*, N1, N5 *(Micrococcus)*, B33, D3, F116, PS4, SD1 *(Pseudomonas)*, NM_1, NT2, 16-6-12, 317 *(Rhizobium)*, P11–M15, 77, 107, 187 *(Staphylococcus)*, 24 *(Streptococcus)*, and A, A1, Leo, P-a-1, R1, VP5, φAG8010, φC, 119 (actinophages).

(b) Heads elongated: ZG/3A (enterobacteria), PT11 *(Agrobacterium)*, BLE, *mor*1, type F *(Bacillus)*, F1 *(Clostridium)*, XP-12 *(Pseudomonas-Xanthomonas)*, 7-7-7 *(Rhizobium)*, 3A *(Staphylococcus)*, VD13, 3ML *(Streptococcus)*, and MSP8, M1, R1, R2 (actinophages).

Family	Phages with short tails	***PODOVIRIDAE***
Main characteristics	Tail short (about 20 nm), noncontractile	
Genus	T7 phage group	–
Type species	Coliphage T7	–
Main characteristics (type species only)	A. Properties of the Virus Particle Nucleic acid: MW $\simeq 25 \times 10^6$; 51% by weight of particle; contains 50% G+C and is nonpermuted and terminally redundant. Protein: About 12 proteins, MW $\simeq 14–150 \times 10^3$; about 450 copies of major capsid protein (MW = 38×10^3); 1 or 2 proteins are located inside the head. Physicochemical properties: MW $\simeq 47 \times 10^6$; $S_{20w} = 507$; buoyant density in CsCl = 1.50 g/cm^3. Infectivity is ether- and chloroform-resistant. Morphology: Isometric head about 65 nm in diameter; short tail (17 nm long) with 6 short fibers. B. Replication Adsorption site is cell wall. Virulent infection. Host chromosome breaks down and viral DNA replicates as concatemer.	

Taxonomic status	*English vernacular name*	*International name*
	C. Biological Aspects Host range: Female enterobacteria.	
Other members	H, PTB, R, T3, Y, W31, φI, φII	

Other members of family include the following viruses, which will probably be assigned to several genera:

(a) Heads isometric: N4, P22, sd, Ω8, 7480b (enterobacteria), PIIBNV6-C *(Agrobacterium)*, Tb *(Brucella)*, HM2 *(Clostridium)*, C1 *(Micrococcus)*, gh-1 *(Pseudomonas)*, φ2042, 2 *(Rhizobium)*, φ17 *(Streptomyces)*, and 114 *(Thermomonospora)*.

(b) Heads elongated: 7-11 (enterobacteria), GA-1, φ29 *(Bacillus)*, 182 *(Streptococcus)*.

Derivation of names
myo: from Greek *mys, myos,* 'muscle', relating to contractile tail.
podo: from Greek *pous, podos,* 'foot', from short tail
Proposed name: stylo: from Greek *stylos,* 'column', from length of tail

References

Ackermann, H.-W.: La classification des bactériophages de *Bacillus* et *Clostridium*. Path. Biol., Paris *22:* 909–917 (1974).

Ackermann, H.-W.: La classification des bactériophages des cocci Gram-positifs: *Micrococcus, Staphylococcus* et *Streptococcus*. Path. Biol., Paris *23:* 247–253 (1975).

Ackermann, H.-W.: La classification des phages caudés des entérobactéries. Path. Biol., Paris *24:* 359–370 (1976).

Ackermann, H.-W.: La classification des bactériophages d'*Agrobacterium* et *Rhizobium*. Path. Biol., Paris *26:* 507–512 (1978).

Ackermann, H.-W.: Tailed bacteriophages; in Laskin, Lechevalier, CRC handbook of microbiology; 2nd ed., vol. 2, pp. 643–671 (CRC Press, West Palm Beach 1978).

Ackermann, H.-W.; Simon, F.; Verger, J.-M.: A survey of *Brucella* phages and morphology of new isolates. Intervirology *16:* 1–7 (1981).

Berthiaume, L.; Ackermann, H.-W.: La classification des actinophages. Path. Biol., Paris *25:* 195–201 (1977).

Liss, A.; Ackermann, H.-W.; Mayer, L.W.; Zierdt, C.H.: Tailed phages of *Pseudomonas* and related bacteria. Intervirology *15:* 71–79 (1981).

Reanney, D.C.; Ackermann, H.-W.: An updated survey of *Bacillus* phages. Intervirology *15:* 190–197 (1981).

Taxonomic status	English vernacular name	International name
Family	–	***PARVOVIRIDAE***

Main characteristics

A. Properties of the Virus Particle

Nucleic acid: Single molecule of ssDNA of MW $1.5–2.0 \times 10^6$. G+C content 41–53%. In some genera the single strands from the virions are complementary and after extraction come together in vitro to form a double strand.

Proteins: Three polypeptides can usually be demonstrated in mature virions of the genera *Parvovirus* and *Dependovirus,* which probably are all derived from a common sequence. Densoviruses seem to have four structural polypeptides. 62–72 protein molecules account for 63–83% of the weight of the virions. Enzymes are probably lacking.

Lipids: None.

Carbohydrate: None.

Physicochemical properties: $MW = 5.5–6.2 \times 10^6$; $S_{20w} = 110–122$. Mature virions band at a density of 1.39–1.42 g/cm^3 in CsCl. Particles with densities >1.42 g/cm^3 are probably 'dense' precursor particles. The infectious particle is stable at pH 3–9, at 56° for at least 60 min, as well as in the presence of lipid solvents. It is inactivated by formalin, β-propiolactone, hydroxylamine, oxidizing agents, and UV irradiation.

Morphology: Nonenveloped isometric particle, 18–26 nm in diameter, with icosahedral symmetry and probably 32 capsomers, 3–4 nm in diameter. Cores measure 14–17 nm.

Antigenic properties: The polypeptides of the virion are immunologically distinguishable; they are, however, antigenically related. In general, antisera to polypeptides do not show neutralization or react with whole virion using HI, complement-fixation or immune electrophoresis. For the genus *Parvovirus* hemagglutinating, CF, and neutralizing antigens are type-specific without cross-reactions, except for minor cross-reactions between rat virus, H-1 and MVM in the fluorescent antibody test. All dependoviruses share a common antigen, as can be demonstrated by fluorescent antibody.

B. Replication

Viral replication takes place in the nucleus where viral proteins in the form of both empty viral capsid structures and progeny infectious virions accumulate. For multiplication, members of the genus *Parvovirus* require one or more cellular functions generated during late S or in early G2 phase of the cell cycle. Members of the genus *Dependovirus* require helper functions (adenoviruses, herpesviruses) for measurable replication.

Genera	Parvovirus group Adeno-associated virus (AAV) Insect parvovirus group	*Parvovirus* *Dependovirus* *Densovirus*

Genus	Parvovirus group	*Parvovirus*
Type species	Parvovirus r-1 (rat virus, Kilham)	–

Main characteristics

A. Properties of the Virus Particle

Nucleic acid: The linear molecule of ssDNA has hairpin structures at both the 5′ and 3′ ends. In most members of the genus all mature virions contain minus-strand DNA. In other members, plus-strand DNA is also incorporated in variable (1–50%) proportion.

Effects on cells: The virion contains a hemagglutinin in most members which has different activities for a variety of red blood cells.

Taxonomic status	*English vernacular name*	*International name*
	B. Replication The virus multiplies in the nucleus, and replication is dependent upon certain functions of the host cell. In consequence, viruses multiply preferentially in actively dividing cells. Formation of intranuclear inclusion bodies. Nucleoli always present, but slightly swollen. C. Biological Aspects Host range: In nature: cat, cattle, dog, goose, mink, raccoon, mouse, pig, rabbit, rat. Under experimental conditions the host range may be extended to homologous or closely related hosts. Rodent viruses and LuIII also replicate in Syrian hamsters. Transmission: Transplacental transmission has been detected for a number of species. Vertical passage by ova is indicated for goose parvovirus. Transmission by mechanical vectors is also possible.	
Other members	Aleutian mink disease Bovine parvovirus Feline parvovirus [subspecies: FPLV (feline panleukopenia virus), MEV (mink enteritis virus), CPV (canine parvovirus)] Goose parvovirus H-1 Lapine parvovirus LuIII Minute of mice (MVM) Porcine parvovirus RT TVX	
Possible members	Gastroenteritis of man (Norwalk agent and related isolates)	
Genus	Adeno-associated virus (AAV)	*Dependovirus*
Type species	Adeno-associated virus type 1	–
Main characteristics	A. Properties of the Virus Particle Nucleic acid: Mature virions contain either positive or negative DNA strands, and the linear single strands come together in vitro to form a double strand. Antigenic properties: All AAVs share a common antigen demonstrable by fluorescent antibody techniques. Replication is dependent upon helper adenoviruses, but infectious AAV DNA, viral antigens and, under certain specific conditions, progeny virions are also made in the presence of herpesviruses.	
Other members	AAV type 2 AAV type 3 AAV type 4 Avian AAV Bovine AAV Canine AAV	

Taxonomic status	English vernacular name	International name
Probable members	Equine AAV Ovine AAV	
Genus	Insect parvovirus group	*Densovirus*
Type species	Densovirus of *Galleria mellonella* (Lepidoptera)	–
Main characteristics	A. Properties of the Virus Particle Nucleic acid: Single strands in virions are either positive or negative, are complementary and come together in vitro to form a double strand. B. Replication Multiply in most of the tissues of larvae, nymphs, and adults without helper virus. Cellular changes consist of hypertrophy of the nucleus with accumulation of virions to form dense voluminous intranuclear masses. C. Biological Aspects Host range: Lepidoptera and probably Diptera and Orthoptera	
Other member	Densovirus of *Junonia*	
Probable members	Densovirus of *Acheta* Densovirus of *Aedes* Densovirus of *Bombyx* Densovirus of *Diatraea* Densovirus of *Nymphalidae* (Agraulis ?) Densovirus of *Sibine*	
Derivation of names	parvo: from Latin *parvus,* 'small' adeno: from Greek *aden, adenos,* 'gland' dependo: from Latin *dependere,* 'depending' denso: from Latin *densus,* 'thick, compact'	

References

Bachmann, P.A.; Hoggan, M.D.; Kurstak, E.; Melnick, J.L.; Pereira, H.G.; Tattersall, P.; Vago, C.: Parvoviridae: second report. Intervirology *11:* 248–254 (1979).

Bloom, M.E.; Race, R.E.; Wolfinbarger, J.B.: Characterization of Aleutian disease virus as a parvovirus. J. Virol. *35:* 836–843 (1980).

Buller, R.M.L.; Janik, J.E.; Sebring, E.D.; Rose, J.A.: Herpes simplex virus types 1 and 2 completely help adenovirus-associated virus replication. J. Virol. *40:* 241–247 (1981).

Clarke, J.K.; McFerran, J.B.; McKillop, E.R.; Curran, W.L.: Isolation of an adeno-associated virus from sheep. Archs Virol. *60:* 171–176 (1979).

Kelly, D.C.; Ayres, M.D.; Spencer, L.K.; Rivers, C.F.: Densonucleosis virus 3: a recent insect parvovirus isolate from *Agraulis vanillae* (Lepidoptera: Nymphalidae). Microbiologica *3:* 455–460 (1980).

Kurstak, E.: Small DNA densonucleosis virus (DNC). Adv. Virus Res. *17:* 207–241 (1972).

Matsunaga, Y.; Matsuno, S.; Mukoyama, J.: Isolation and characterization of a parvovirus of rabbits. Infect. Immunity *18:* 495–500 (1977).

McMaster, G.K.; Tratschin, J.D.; Siegl, G.: Comparison of canine parvovirus with mink enteritis virus by restriction site mapping. J. Virol. *38:* 368–371 (1981).
Moore, N.F.; Kelly, D.C.: Interrelationships of the proteins of two insect parvoviruses (densonucleosis virus types 1 and 2). Intervirology *14:* 160–166 (1980).
Siegl, G.: The parvoviruses. Virology Monographs, vol. 15 (Springer, Vienna 1976).
Ward, D.; Tattersall, P. (eds.): Replication of mammalian parvoviruses (Cold Spring Harbor Press, Cold Spring Harbor 1978).

Taxonomic status	*English vernacular name*	*International name*
Group	–	***GEMINIVIRUS***
Type member	Maize streak virus (MSV) (133)	–
Main characteristics	A. Properties of the Virus Particle Nucleic acid: One molecule of circular, positive-sense ssDNA, MW $7–8 \times 10^5$. Protein: Single coat polypeptide, MW $28–34 \times 10^3$. Lipid: None reported. Carbohydrate: None reported. Physicochemical properties: $S_{20w} \simeq 70$ (for particle pairs). Morphology: Geminate particles, 18×20 nm, consisting of two incomplete icosahedra wit T=1 surface lattice with a total of 22 capsomers. Antigenic properties: Efficient immunogens. Single precipitin line in gel-diffusion tests. B. Replication Virus particles accumulate in nucleus producing large aggregates. C. Biological Aspects Host range: Wide variety of plants infected by members, but each member has narrow hos range. Transmission: Some members transmitted by leafhoppers and others by whiteflies in persistent manner. Some members transmitted experimentally by mechanical inoculation	
Other members	Bean golden mosaic (192) Cassava latent *Chloris* striate mosaic Tomato golden mosaic (possibly synonymous with tomato yellow mosaic)	
Probable members	Beet curly top (210) *Euphorbia* mosaic Mungbean yellow mosaic *Paspalum* striate mosaic Tobacco leafcurl (= tomato yellow dwarf) Tobacco yellow dwarf (= bean summer death) Tomato yellow leafcurl Tomato yellow mosaic Wheat dwarf	
Derivation of name	gemini: from Latin *gemini,* 'twins', from characteristic double particle	

References

Goodman, R.M.: Geminiviruses; in Kurstak, Handbook of plant virus infections and comparative diagnosis pp. 879–910 (Elsevier/North Holland, Amsterdam 1981a).

Goodman, R.M.: Geminiviruses. J. gen. Virol. *54:* 9–21 (1981b).

Hamilton, W.D.O.; Saunders, R.C.; Coutts, R.H.A.; Buck, K.W.: Characterization of tomato golden mosai virus as a geminivirus. FEMS Microbiol. Lett. *11:* 263–267 (1981).

Taxonomic status	*English vernacular name*	*International name*
Family	ϕX phage group	***MICROVIRIDAE***
Genus	ϕX phage group	*Microvirus*
Type species	Phage ϕX174	–

Main characteristics A. Properties of the Virus Particle

Nucleic acid: One piece of circular ssDNA; MW$\simeq 1.7 \times 10^6$ (26% by weight of particle); 35–60% G+C content.

Protein: 60 molecules of capsid protein (MW = 50,000) and at least 3 other proteins.

Lipid: None.

Carbohydrate: None.

Physicochemical properties: MW$\simeq 6.7 \times 10^6$; $S_{20w} = 114$; buoyant density in CsCl$\simeq 1.40$ g/cm^3.

Morphology: Virions are icosahedral and about 27 nm in diameter. They have 12 capsomers (T = 1) and knob-like spikes on vertices. No envelope.

B. Replication

Adsorption to cell wall. Phage DNA is converted to a circular dsRF which replicates in a semiconservative way. Progeny ssDNA is generated by displacement from RF DNA. Genome codes for 9 major proteins, most of them involved in DNA replication. Two sections of the genome code for two different proteins in different reading frames. No inclusion bodies; lysis.

C. Biological Aspects

Host range: Enterobacteria.

Other members dϕ3, dϕ4, dϕ5, G4, G6, G13, 1ϕ1, 1ϕ3, 1ϕ7, 1ϕ9, M20, St-1, S13, WA/1, WF/1, α3, α10, δ1, ζ3, η8, o6, ϕA, ϕR, U3

Derivation of name micro: from Greek *micros*, 'small'

References

Ackermann, H.-W.: Cubic, filamentous, and pleomorphic bacteriophages; in Laskin, Lechevalier, CRC handbook of microbiology; 2nd ed., vol. II, pp. 673–682 (CRC Press, West Palm Beach 1978).

Denhardt, D.T.: The isometric single-stranded DNA phages; in Fraenkel-Conrat, Wagner, Comprehensive virology, vol. 7, pp. 1–104 (Plenum, New York 1977).

Taxonomic status	English vernacular name	International name
Family	Rod-shaped phages	***INOVIRIDAE***
Main characteristics	A. Properties of the Virus Particle Nucleic acid: One piece of circular ssDNA. Physicochemical properties: Infectivity is chloroform-sensitive and heat-resistant. Morphology: Virions are long or short rods. Capsids are helical and seem to be tubules. Particles of abnormal length are frequent. No envelope. B. Replication Infecting viral DNA is converted into a dsRF which replicates in a semiconservative way. No inclusion bodies. Phages extruded through host membranes, no lysis, host survives.	
Genera	Filamentous phages Mycoplasma virus type 1 phages	*Inovirus* *Plectrovirus*
Genus	Filamentous phages	*Inovirus*
Type member (proposed)	fd	–
Main characteristics	A. Properties of the Virus Particle Nucleic acid: MW = $1.9–2.7 \times 10^6$; 5.5–12% by weight of particle; G + C content 42–62%. Protein: One major coat protein (MW $\simeq 5 \times 10^3$) and 3 or 4 copies of maturation protei (MW = $65–70 \times 10^3$). Coat proteins appear to lack cysteine and histidine. Lipid: None. Carbohydrate: None. Physicochemical properties: Particle MW = $11–23 \times 10^6$; $S_{20w} = 41–45$; buoyant density i CsCl $\simeq 1.29$ g/cm^3. Infectivity is sensitive to sonication; ether sensitivity is variable. Morphology: Flexible rods, 760–1,950 nm long × 6 nm diameter. B. Replication Virions adsorb slowly to pili or poles(?) of bacteria and enter the cells. Most are specifi for male bacteria. Progeny viral ssDNA is produced by displacement from RF DNA. Genome consists of at least 8 genes. Mature virions appear to be assembled at the ce membrane as the phage leaves the cell. C. Biological Aspects Host range: Enterobacteria, *Pseudomonas, Vibrio, Xanthomonas.*	
Possible members	Subgroups based on length, host range, serological relationships, chemical compositio (a) fd group: AE2, Ec9, f1, HR, M13, ZG/2, ZJ/2, δA (about 800 nm, enterobacteria possibly Pf3 (760 nm, enterobacteria and *Pseudomonas*). (b) If1, If2, IKe? (1,300 nm, enterobacteria). (c) Pf1, Pf2 (1,915 nm, *Pseudomonas*). (d) Cf, Xf, Xf2 (980 nm, *Xanthomonas*). (e) v6 (*Vibrio*).	
Genus	Mycoplasma virus type 1 phages	*Plectrovirus*

Taxonomic status	*English vernacular name*	*International name*
Type member (proposed)	MVL51	–
Main characteristics	A. Properties of the Virus Particle Nucleic acid: MW = 1.5×10^6. Protein: Four proteins (MW $19–70 \times 10^3$). Physicochemical properties: $S_{20w} = 40–100$; buoyant density in CsCl = 1.37 g/cm³. Infectivity is ether- and detergent-resistant. Morphology: Virions are short, straight rods of about 84×14 nm, have one rounded end, and may be derived from icosahedra (T = 1). B. Replication See Family 'Main characteristics' C. Biological Aspects Host range: *Acholeplasma*.	
Other members	MV-L1, MVG51, O3c1r, 10tur, others (over 30 isolates)	
Possible member	SV-C1 ($230–280 \times 10–15$ nm, host *Spiroplasma*)	
Derivation of name	ino: from Greek *is, inos,* 'muscle'	

References

Ackermann, H.-W.: Cubic, filamentous, and pleomorphic bacteriophages; in Laskin, Lechevalier, CRC handbook of microbiology; 2nd ed., vol. II, pp. 673–682 (CRC Press, West Palm Beach 1978).

Maniloff, J.; Das, J.; Christensen, J. R.: Viruses of mycoplasmas and spiroplasmas. Adv. Virus Res. *21:* 343–380 (1977).

Ray, D. S.: Replication of filamentous bacteriophages; in Fraenkel-Conrat, Wagner, Comprehensive virology, vol. 7, pp. 105–178 (Plenum, New York 1977).

Taxonomic status	*English vernacular name*	*International name*
Family	ϕ6 phage group	***CYSTOVIRIDAE***
Genus (monotypic)	ϕ6 phage group	*Cystovirus*
Type species	Phage ϕ6	–

Main characteristics

A. Properties of the Virus Particle
Nucleic acid: Three pieces of linear dsRNA; total MW$\simeq$10.4 × 10^6 (2.3, 3.1 and 5.0 × 10^6); 10% by weight of particle; 56% G+C content.
Protein: Eleven polypeptides with total MW = 364 × 10^3 (range 6–82 × 10^3); transcriptase activity present.
Lipid: Located in the envelope, constitutes about 20% of particle; over 90% is phospholipid.
Physicochemical properties: MW$\simeq$100 × 10^6; S_{20w} = 446; buoyant density in CsCl = 1.27 g/cm^3. Infectivity is ether-, chloroform-, and detergent-sensitive.
Morphology: Isometric, about 75 nm; flexible envelope and cubic capsid of 60 nm diameter.

B. Replication
Adsorption to sides of pili; no inclusion bodies; lysis.

C. Biological Aspects
Host range: *Pseudomonas*

Derivation of name: cysto: from Greek *kystis,* 'bladder, sack'

References

Ackermann, H.-W.: Cubic, filamentous, and pleomorphic bacteriophages; in Laskin, Lechevalier, CRC handbook of microbiology; 2nd ed., vol. II, pp. 673–682 (CRC Press, West Palm Beach 1978).

Day, L. A.; Mindich, L.: The molecular weight of bacteriophage ϕ6 and its nucleocapsid. Virology *103:* 376–385 (1980).

Mindich, L.: Bacteriophages that contain lipid; in Fraenkel-Conrat, Wagner, Comprehensive virology, vol. 12, pp. 271–335 (Plenum, New York 1978).

Taxonomic status	*English vernacular name*	*International name*
Family	–	***REOVIRIDAE***

Main characteristics

A. Properties of the Virus Particle

Nucleic acid: 10–12 pieces of linear dsRNA; MWs$\simeq$0.2–3.0 $\times 10^6$. Total MW = 12–20 $\times 10^6$. About 14–22% by weight of virus.

Protein: 6–10 polypeptides in virion, including transcriptase and other enzymes. MWs$\simeq$ 15–155 $\times 10^3$.

Lipid: None.

Carbohydrate: Some polypeptides may contain a small amount of carbohydrate.

Physicochemical properties: MW of virion$\simeq$65–200 $\times 10^6$. Effective buoyant density in CsCl = 1.36–1.39 g/cm^3.

Morphology: Icosahedral particle with diameter$\simeq$60–80 nm; no lipoprotein envelope; two protein coats; particle with the outer coat removed is termed the core; transcriptase activity associated with the core. Cores have 12 spikes with 5-fold symmetry arranged icosahedrally. Genus *Fijivirus* also has spikes on virions.

B. Replication

In cytoplasm. Viroplasms in cytoplasm of infected cells, sometimes containing virus particles in crystalline arrays. Genetic recombination has been demonstrated with some genera, where it occurs very efficiently by genome segment reassortment.

Taxonomic status	English vernacular name	International name
Genera	Reovirus subgroup	*Reovirus*
	–	*Orbivirus*
	–	*Rotavirus*
	Plant reovirus subgroup 1	*Phytoreovirus*
	Plant reovirus subgroup 2	*Fijivirus*
	Cytoplasmic polyhedrosis virus group	–
	Other members, ungrouped	–
Genus	Reovirus subgroup	*Reovirus*
Type species	Reovirus type 1	–

Main characteristics

A. Properties of the Virus Particle

Nucleic acid: 10 pieces with MWs = 0.5–2.7 $\times 10^6$; total MW = 14–15 $\times 10^6$. 14% by weight of virus; 42–44% G + C. Virus contains about 3,000 molecules of ss oligoribonucleotides 2–20 nucleotides long. There is no sequence homology between *Reovirus* members and members of other genera.

Protein: Nine polypeptides with MWs = 34–155 $\times 10^3$; 86% of virus by weight. Nucleotide phosphohydrolase and capping enzymes present besides the transcriptase, which requires proteolytic enzymes for activation.

Physicochemical properties: MW = 127–131 $\times 10^6$. S_{20w} = 734. Infectivity resistant to ether; relatively heat-stable; stable at pH 3.0.

Morphology: 76 nm diameter. Cores, 52 nm diameter, contain about 45% RNA.

Antigenic properties: The type-specific antigen is protein σ_1; serological types 1, 2 and 3 cross-react.

Taxonomic status	*English vernacular name*	*International name*
	B. Replication Two nonstructural proteins are synthesized (MWs = 36,000 and 80,000). Transcriptase synthesizes positive strands. Later a presumably related replicase makes negative strands, thus forming ds progeny RNA molecules. C. Biological Aspects Host range: Vertebrates only; man, monkeys, birds, cattle, bats. Experimentally in cells of most vertebrate species. Transmission: Horizontal.	
Other members	Serologic types 1, 2 and 3 include strains isolated from man, monkeys, dogs and cattle. Avian strains include Crawley, Nelson Bay.	
Genus	–	*Orbivirus*
Type species	Bluetongue virus	–
Main characteristics	A. Properties of the Virus Particle Nucleic acid: 10 pieces with MWs = $0.3–2.7 \times 10^6$; total MW $\simeq 12 \times 10^6$. 20% by weight of virus; 42–44% G+C. Nucleic acid hybridization analysis between bluetongue virus subgroups indicates that some genome segments hybridize very extensively, others very poorly. Protein: Seven polypeptides with MWs = $32–155 \times 10^3$. 80% by weight of virus. Removal of outer shell required for activation of the RNA-dependent RNA polymerase. Physicochemical properties: Infectivity lost at pH 3.0. Lipid solvents reduce infectivity about 10-fold. MW $\simeq 60 \times 10^6$. $S_{20w} = 550$; 65–80 nm diameter (68 nm for bluetongue virus). The inner shell has 32 morphological units showing circular surface configurations. These 32 units are visible when the outer shell is still present, a distinguishing feature of the genus *Orbivirus*. Antigenic properties: Polypeptide 2 is the main antigenic determinant for neutralization. B. Replication Two nonstructural proteins with MWs = 40,000 and 50,000. Replication probably like that of *Reovirus* takes place in cytoplasmic viroplasms. Morphogenesis is accompanied by formation of regularly structured filaments and tubules. C. Biological Aspects Host range: Insects and other arthropods. Vertebrate species including man, horses, monkeys, rabbits, cattle, deer, suckling mice. Transmission: Vectors: culicoides, mosquitoes, phlebotomines and ticks.	
Other members	There are some 12 serological subgroups of genus *Orbivirus*. (Vectors are indicated where known.) They are: (a) Bluetongue subgroup Bluetongue virus (21 serotypes) (culicoides) (b) Eubenangee subgroup Eubenangee (mosquitoes) Pata (mosquitoes) Tilligerry (NB 7080) (mosquitoes)	

Taxonomic status	*English vernacular name*	*International name*
	(c) Corriparta subgroup	
	Acado	(mosquitoes)
	Bambari	
	Corriparta	(mosquitoes)
	Jacareacanga	
	(d) Changuinola subgroup	
	Be Ar 35646	(phlebotomines)
	Be Ar 41067	(phlebotomines)
	Be Ar 54342	(phlebotomines)
	Changuinola	(phlebotomines)
	Irituia	
	(e) Colorado tick fever subgroup	
	Colorado tick fever	(ticks)
	Eyach	(ticks)
	(f) Kemerovo subgroup	
	Baku	(ticks)
	Bauline	(ticks)
	Cape Wrath	(ticks)
	Chenuda	(ticks)
	Fin isolate	
	Great Island	(ticks)
	Huacho	(ticks)
	Kemerovo	(ticks)
	Kenai	
	Lipovnik	(ticks)
	Mono Lake	(ticks)
	Mykenes	(ticks)
	Nugget	(ticks)
	Okhotskiy	(ticks)
	Poovoot	
	Seletar	(ticks)
	Sixgun City	(ticks)
	Tindholmur	(ticks)
	Tribec	(ticks)
	Wad Medani	(ticks)
	Yaquina Head	(ticks)
	(g) Palyam subgroup	
	Abadina	(culicoides)
	D'Aguilar	(culicoides)
	Kasba	(mosquitoes)
	Nyabira	
	Palyam	(mosquitoes)
	Vellore	(mosquitoes)
	(h) Epizootic disease of deer subgroup	
	EHD, New Jersey	
	EHD, Can Alberta	

Taxonomic status	English vernacular name	International name
	Ib Ar 22619	
	Ib Ar 33853	
	Ibaraki	
	(i) Warrego subgroup	
	Mitchell River (MRM 10434)	(culicoides)
	Warrego (Ch 9935)	(culicoides)
	(j) Wallal subgroup	
	Mudginbarry	
	Wallal (CH 12048)	
	(k) African horse sickness (9 serotypes)	
	(l) Equine encephalosis (5 serotypes)	
	(m) Ungrouped	
	Ife	
	Japanant	
	Lebombo	
	Llano Seco	
	Orunga	
	Paroo River	
	T-50616 (skunk isolate)	
	Umatilla	
Probable member	Rabbit syncytium virus	
Genus	–	*Rotavirus*
Type species	Human rotavirus	–
Main characteristics	A. Properties of the Virus Particle Nucleic acid: 11 pieces with MWs = $0.2–2.2 \times 10^6$. Protein: 8–10 polypeptides with MWs = $15–130 \times 10^3$. Physicochemical properties: $S_{20w} = 525$. Infectivity stable at pH 3.0 and relatively heat-stable. Resistant to ether. Morphology: Icosahedral (P=3; T=3). Diameter 65–75 nm. Capsomers composed of shared subunits. B. Replication In the cytoplasm. C. Biological Aspects Host range: Mammals. Disease is caused by homologous virus in humans, mice, calves, piglets, foals and lambs. Human rotavirus infections have been demonstrated in monkeys, piglets and calves. Transmission: Horizontal. No vectors.	

Taxonomic status	*English vernacular name*	*International name*
Other members	Rotaviruses have been isolated from man, cattle, mouse (EDIM), guinea pig, sheep, goat, pig, monkey (SA11), horse, antelope, bison, deer, rabbit, dog and duck. 'O' agent may have originated in cattle or sheep. Serotypes are probably numerous and cross-react.	
Genus	Plant reovirus subgroup 1	*Phytoreovirus*
Type species	Wound tumor virus (WTV) (34)	–

Main characteristics

A. Properties of the Virus Particle

Nucleic acid: 12 pieces with MWs$\simeq$0.3–3.0 $\times 10^6$, with total MW$\simeq$16 $\times 10^6$. 22% by weight of virus. 38–44% G+C.

Protein: Seven polypeptides with MWs = 35–160 $\times 10^3$. 78% by weight of virus. Removal of outer shell not required for activation of the transcriptase.

Physicochemical properties: MW$\simeq$65 $\times 10^6$. S_{20w} = 510. Optimal stability at pH 6.6. Resistant to Freon and CCl_4.

Morphology: Diameter$\simeq$70 nm. Icosahedral, rather angular shape. WTV possesses an outer amorphous layer (2 polypeptides), an outer layer of distinct capsomers, and a smooth core (3 polypeptides) of MWs = 58, 118, and 160 $\times 10^3$. The core is about 59 nm in diameter with no spikes.

B. Replication

In cytoplasmic viroplasms. Messenger RNA transcripts have the 5′ terminal sequence: $m^7G^{5'}ppp^{5'}Pp$... WTV replicates in both plant and leafhopper hosts. Continuous propagation in plants without access to vectors can lead to the selection of mutants which lack some genome segments and which may no longer replicate in the insect.

C. Biological Aspects

Host range: In nature WTV was found only once, in the leafhopper *Agalliopsis novella*. No natural plant host known. Natural hosts of rice dwarf virus (RDV) are cicadellid hoppers and rice. Experimental host range of WTV is wide among dicotyledonous plants. RDV has a narrow host range among the *Gramineae*; WTV grows in cell lines derived from embryonic tissues of vectors.

Transmission: Only by cicadellid leafhoppers *(Agallia, Agalliopsis, Nephotettix)*. Transmission is propagative; acquisition after 1 min or more; latent period about 2 weeks, then lifelong transmission by insects to plants. Transovarial in insect vectors. Transmitted to insects by injection into hemocoele.

Other member	Rice dwarf virus (RDV) (102)	
Possible member	Rice gall dwarf virus	
Genus	Plant reovirus subgroup 2	*Fijivirus*
Type species	Fiji disease virus (FDV) (119)	–

Main characteristics

A. Properties of the Virus Particle

Nucleic acid: 10 pieces with MWs$\simeq$1.0–2.9 $\times 10^6$ with total MW = 18–20 $\times 10^6$; about 45% G+C.

Taxonomic status	English vernacular name	International name
	Protein: Not known for FDV. For maize rough dwarf virus (MRDV), 7 polypeptides wit MWs = 64–139 × 10^3. Physicochemical properties: Not established. Morphology: External appearance is spherical with diameter = 65–71 nm (in uranyl ace tate). 12 external knobs ≃ 11 nm in diameter and 8–16 nm long (A spikes) located one o each 5-fold axis. The structure breaks down spontaneously in vitro to give spiked core 54 nm in diameter, which have 12 icosahedrally located spikes (B spikes, ≃ 8 nm high, 14–1 nm wide). Treatment of MRDV with various reagents produces smooth (spikeless) cores 50–55 nm in diameter and containing 3 polypeptides, with MWs = 123, 126 and 139 × 10^3. B. Replication In cytoplasmic viroplasms, consisting of a matrix containing filaments. C. Biological Aspects Host range: Flowering plants: confined to *Gramineae*. Insects: confined to plant hopper *(Delphacidae, Auchenorhyncha, Hemiptera)*. Transmission: In nature only by Delphacid plant hoppers, e.g., *Laodelphax, Javesella Delphacodes, Sogatella, Perkinsiella, Unkanodes*. Transmission is propagative; acquisitio after some hours feeding; latent period about 2 weeks; then lifelong transmission by th insect to plants. FDV can be transmitted transovarially, for other members, conflicting reports. Mechanical transmission into insects, with high efficiency by injection into hemo coele.	
Other members	The following are members of this genus based upon morphology, demonstrated possessio of dsRNA, and antigenic properties; not all of these have been confirmed for all. A serologi grouping which is as yet provisional is: Group I – Cereal tillering disease, Maize rough dwarf (72), Pangola stunt (175), Rice black streaked dwarf (135) — serologically related Group II – Fiji disease (type species) (119) Group III – *Arrhenatherum* blue dwarf, *Lolium* enation, Oat sterile dwarf (217) — serologically related	
Genus	Cytoplasmic polyhedrosis virus group	–
Type species	Cytoplasmic polyhedrosis virus from *Bombyx mori*	–
Main characteristics	A. Properties of the Virus Particle Nucleic acid: 10 pieces with MWs = 0.3–2.7 × 10^6, with total MW = 13–16 × 10^6; 25–30% by weight of virus. 36–42% G + C. Segments do not hybridize with each other and have n homology with members of other genera. The positive strands of the virion RNA ar methylated and capped at the 5′ terminus. Protein: Three to five polypeptides of MWs 30–151 × 10^3. 70–75% by weight of virus Transcriptase in particle does not require treatment with proteolytic enzyme for activatio Also present: nucleotide phosphohydrolase; capping enzymes; exonuclease; hemagglutini for chicken, sheep, and mouse erythrocytes.	

Taxonomic status	*English vernacular name*	*International name*
	Physicochemical properties: MW$\simeq$50 × 10^6. S_{20w}=370–440. Infectivity: stable at pH 3.0; lost after 10 min at 80–85°; resistant to ether. Relatively stable to UV irradiation. Capsid resistant to proteolytic enzymes such as chymotrypsin. Morphology: Spherical, 50–65 nm diameter. Twelve apparently hollow spikes located at icosahedral vertices. Dense core area surrounded by an outer shell, but no clearly defined outer capsid structure like that of *Reovirus*. B. Replication Proteins usually accumulate in the cytoplasm, with a pronounced cellular tropism for midgut epithelial cells. Many virus particles are occluded within 'polyhedra' composed of one major polypeptide, with MW=25–30 × 10^3. Carbohydrate is associated with this protein. Virus-specific ssRNA is synthesized during infection, probably by the transcriptase found in the virion. Some viral RNA synthesis may occur in the nucleus. Replication occurs in insect cell cultures. C. Biological Aspects Host range: Insects: Lepidoptera, Diptera, Hymenoptera. Crustacea: Simocephalus. Transmission: Horizontal.	
Other members	Eleven 'types' defined by the distinctive electrophoretic profiles of their RNA genome segments (in addition to type 1, the type species): Type 2 from *Inachis io* Type 3 from *Spodoptera exempta* Type 4 from *Actias selene* Type 5 from *Trichoplusia ni* Type 6 from *Biston betularia* Type 7 from *Triphena pronuba* Type 8 from *Abraxas grossulariata* Type 9 from *Agrotis segetum* Type 10 from *Aporophylla lutulenta* Type 11 from *Spodoptera exigua* Type 12 from *Spodoptera exempta*	
Probable members	Viruses from approximately 150 different insect species	
Genus	Other members, but so far ungrouped. Leafhopper A virus (LAV) (possibly involved in maize wallaby ear disease) similar to *Fijivirus*, but replicates in Cicadellid not Delphacid insects; may not replicate in plants. Rice ragged stunt virus, resembles *Fijivirus*, but has a different genome (8 pieces with MWs =0.5–2.5 × 10^6 with total MW$\simeq$12 × 10^6.	
Derivation of names	reo: sigla from *r*espiratory *e*nteric *o*rphan orbi: from Latin *orbis*, 'ring' rota: from Latin *rota*, 'wheel' phyto: from Greek *phyton*, 'plant' fiji: from name of country from which virus was first described	

References

Boccardo, G.; Hatta, T.; Francki, R.I.B.; Grivell, C.J.: Purification and some properties of reovirus-like particles from leafhoppers and their possible involvement in wallaby ear disease of maize. Virology *100:* 300–313 (1980).

Boccardo, G.; Milne, R.G.: Electrophoretic fractionation of the double-stranded RNA genome of rice ragged stunt virus. Intervirology *14:* 57–60 (1980).

Boccardo, G.; Milne, R.G.; Luisoni, E.: Purification, serology and nucleic acid of pangola stunt virus subviral particles. J. gen. Virol. *45:* 659–664 (1979).

Farrell, J.A.; Harvey, J.D.; Bellamy, A.R.: Biophysical studies of reovirus type 3. I. The molecular weights of reovirus and reovirus cores. Virology *62:* 145–153 (1974).

Flewett, T.H.; Woode, G.N.: The rotaviruses. Archs Virol. *57:* 1–23 (1978).

Hatta, T.; Francki, R.I.B.: Morphology of Fiji disease virus. Virology *76:* 797–807 (1977).

Joklik, W.K.: The reproduction of Reoviridae; in Fraenkel-Conrat, Wagner, Comprehensive virology, vol. 2, pp. 231–334 (Plenum, New York 1974).

McNulty, M.S.: Rotaviruses. J. gen. Virol. *40:* 1–18 (1978).

Milne, R.G.; Lovisolo, O.: Maize rough dwarf and related viruses. Adv. Virus Res. *21:* 267–341 (1977).

Milne, R.G.; Luisoni, E.: Serological relationships among maize rough dwarf-like viruses. Virology *80:* 12–20 (1977).

Omura, T.; Inoue, H.; Morinaka, T.; Saito, Y.; Chettanachit, D.; Putta, M.; Parejarearn, A.; Disthaporn, S.: Rice gall dwarf disease. Plant Dis. *64:* 795–797 (1980).

Payne, C.C.; Harrap, K.A.: Cytoplasmic polyhedrosis viruses; in Maramorosch, The atlas of insect and plant viruses, pp. 105–129 (Academic Press, New York 1977).

Payne, C.C.; Rivers, C.F.: A provisional classification of cytoplasmic polyhedrosis viruses based on the sizes of the RNA genome segments. J. gen. Virol. *33:* 71–85 (1976).

Reddy, D.V.R.; MacLeod, R.: Polypeptide components of wound tumor virus. Virology *70:* 274–282 (1976).

Shatkin, A.J.; Both, G.W.: Reovirus mRNA: transcription and translation. Cell *7:* 305–313 (1976).

Silverstein, S.C.; Christman, J.K.; Acs, G.: The reovirus replicative cycle. A. Rev. Biochem. *45:* 375–408 (1976).

Taxonomic status	*English vernacular name*	*International name*
None (possible family)	Isometric dsRNA mycoviruses requiring only one RNA segment for replication	–
Main characteristics	A. Properties of the Virus Particle Nucleic acid: Single molecule of dsRNA, MW 3.0–5.7 × 10^6; additional segments of dsRNA (satellite or defective) may be present in some virus isolates. Protein: Single major capsid structural polypeptide species, MW 73–120 × 10^3. Lipid: None. Carbohydrate: None. Physicochemical properties: S_{20w} = 160–283 (particles containing dsRNA). Morphology: Polyhedral, diameter 35–48 nm, single shell of protein. B. Replication In cytoplasm.	
None (possible genera)	*Saccharomyces cerevisiae* virus group *Helminthosporium maydis* virus group	– –
None (possible genus)	*Saccharomyces cerevisiae* virus group	–
None (possible type species)	*Saccharomyces cerevisiae* virus ScV1 (from strain S7)	–
Main characteristics	A. Properties of the Virus Particle Nucleic acid: Single molecule of linear dsRNA, MW 3.0–4.2 × 10^6; additional segments of dsRNA (satellite or defective) present in some isolates. Protein: Single major capsid polypeptide species, MW 73–88 × 10^3. RNA polymerase (transcriptase) present. Physicochemical properties: S_{20w} = 160–172 (for particles containing dsRNA). Morphology: Diameter 35–43 nm. B. Replication The virion-associated RNA polymerase catalyzes in vitro transcription of dsRNA, probably by a conservative mechanism. C. Biological Aspects Transmission: Congenital through mating and sporogenesis.	
Other members	*Saccharomyces cerevisiae* virus ScV2 *Ustilago maydis* virus (from strain 3004)	
Probable member	*Mycogone perniciosa* virus	
Possible members	*Aspergillus foetidus* virus S *Aspergillus niger* virus S *Gaeumannomyces graminis* viruses 3b1a-A and F6-A *Helminthosporium victoriae* virus (HvV-A) *Thielaviopsis basicola* viruses	

Taxonomic status	*English vernacular name*	*International name*
None (possible genus)	*Helminthosporium maydis* virus group (monotypic)	–
None (possible type species)	*Helminthosporium maydis* virus	–
Main characteristics	A. Properties of the Virus Particle Nucleic acid: Single molecule of linear dsRNA, MW 5.7×10^6. Protein: Single major capsid polypeptide species, MW 121×10^3. Physicochemical properties: $S_{20w} = 283$ (for particles containing dsRNA). Morphology: 48 nm in diameter.	

References

Adler, J.P.; Wood, H.A.; Bozarth, R.F.: Virus-like particles from killer, neutral and sensitive strains of *Saccharomyces cerevisiae*. J. Virol. *17:* 472–476 (1976).

Barton, R.J.: *Mycogone perniciosa* virus. Report of the Glasshouse Crops Research Institute for 1977, p. 133 (1978)

Bozarth, R.F.: Biophysical and biochemical characterization of virus-like particles containing a high molecular weight dsRNA from *Helminthosporium maydis*. Virology *80:* 149–157 (1977).

Bozarth, R.F.; Goenaga, A.: A complex of virus-like particles from *Thielavopsis basicola*. J. Virol. *24:* 846–849 (1977).

Bozarth, R.F.; Koltin, Y.; Weissman, M.B.; Parker, R.L.; Dalton, R.E.; Steinlauf, R.: The molecular weight and packaging of dsRNAs in the mycovirus from *Ustilago maydis* killer strains. Virology *113:* 492–502 (1981)

Brennan, V.E.; Field, L.; Cizdziel, P.; Bruenn, J.A.: Sequences at the 3′ ends of yeast viral dsRNAs: proposed transcriptase and replicase initiation sites. Nucl. Acid Res. *9:* 4007–4021 (1981).

Buck, K.W.; Almond, M.R.; McFadden, J.J.P.; Romanos, M.A.; Rawlinson, C.J.: Properties of thirteen viruses and virus variants obtained from eight isolates of the wheat take-all fungus, *Gaeumannomyces graminis* var. *tritici*. J. gen. Virol. *53:* 235–245 (1981).

Buck, K.W.; Lhoas, P.; Border, D.J.; Street, B.K.: Virus particles in yeast. Biochem. Soc. Trans. *1:* 1141–1142 (1973).

Hopper, J.E.; Bostian, K.A.; Rowe, L.B.; Tipper, D.J.: Translation of the L-species dsRNA genome of the killer-associated virus-like particle of *Saccharomyces cerevisiae*. J. biol. Chem. *252:* 9010–9017 (1977).

Koltin, Y.: Virus-like particles in *Ustilago maydis:* mutants with partial genomes. Genetics *86:* 527–534 (1977).

Sanderlin, R.S.; Ghabrial, S.A.: Physicochemical properties of two distinct types of virus-like particles from *Helminthosporium victoriae*. Virology *87:* 142–151 (1978).

Taxonomic status	*English vernacular name*	*International name*
None (possible family)	Isometric dsRNA mycoviruses requiring two RNA segments for replication	–

Main characteristics

A. Properties of the Virus Particle

Nucleic acid: Two segments of monocistronic dsRNA, MW 0.9–1.6 × 10^6 (for individual segments), each separately encapsidated, both required for virus replication; additional segments of dsRNA (satellite or defective) may be present in some virus isolates.
Protein: Single major capsid structural polypeptide species, MW 42–72 × 10^3. RNA polymerase present.
Lipid: None.
Carbohydrate: None.
Physicochemical properties: S_{20w} = 101–145 (for particles containing dsRNA).
Morphology: Polyhedral, diameter 30–35 nm, single shell of protein.

B. Replication
In cytoplasm.

None (possible genera)	*Penicillium stoloniferum* PsV-S group	–
	Gaeumannomyces graminis virus group I	–
	Gaeumannomyces graminis virus group II	–
None (possible genus)	*Penicillium stoloniferum* PsV-S group	–
None (possible type species)	*Penicillium stoloniferum* virus S (PsV-S)	–

Main characteristics

A. Properties of the Virus Particle

Nucleic acid: Two pieces of linear dsRNA, MW 1.10 and 0.94 × 10^6, separately encapsidated, comprising 16% of particle weight. Virus messenger RNAs of half the dsRNA MWs are also separately encapsidated.
Protein: Single major capsid polypeptide species, MW 42 × 10^3; one minor polypeptide species, MW 56 × 10^3, probably the virion RNA polymerase.
Physicochemical properties: Particles of four types (E, M, L and H) with MWs ranging from 5.0 to 6.7 × 10^6; S_{20w} = 66 (E), 87 (M), 101 (L) and 113 (H); buoyant densities (g/cm^3) in CsCl = 1.30 (E), 1.33 (M), 1.36 (L) and 1.39 (H). E particles are empty capsids, M particles contain ssRNA (mRNA), L particles contain dsRNA, and H particles contain dsRNA and ssRNA (probably replicative intermediates). Particles disrupted by sodium dodecyl sulfate.
Morphology: Diameter 30–34 nm.
Antigenic properties: Serological relationships between PsV-S and viruses of *Diplocarpon rosae* and *Aspergillus ochraceous.*

B. Replication
Virus particles occur in cytoplasm, often in regular aggregates in vesicles or vacuoles. The virion-associated RNA polymerase catalyzes in vitro semi-conservative replication of dsRNA.

Taxonomic status	*English vernacular name*	*International name*
	C. Biological Aspects Transmission: Congenital within conidia, and by hyphal anastomosis.	
Probable members	Virus of *Aspergillus ochraceous*, virus of *Diplocarpon rosae*	
Possible member	*Penicillium stoloniferum* virus F (PsV-F)	
None (possible genus)	*Gaeumannomyces graminis* virus group I	–
None (possible type species)	*Gaeumannomyces graminis* virus 019/6-A	–
Main characteristics	A. Properties of the Virus Particles Nucleic acid: Two pieces of linear dsRNA, MW $1.19–1.22 \times 10^6$ and $1.27–1.30 \times 10^6$, separately encapsidated. Some isolates contain additional dsRNA species (satellite or defective). Protein: Single major capsid polypeptide species, MW $54–60 \times 10^3$. RNA polymerase (transcriptase) present. Physicochemical properties: $S_{20w} = 109–128$. Morphology: Diameter 35 nm. Antigenic properties: Serological relationships between members. B. Replication The virion-associated RNA polymerase catalyzes in vitro transcription of dsRNA to produce virus mRNAs, probably by a semi-conservative displacement mechanism. C. Biological Aspects Transmission: Congenital within conidia, and by hyphal anastomosis. Not transmitted in ascospores.	
Other members	*G. graminis* viruses 38-4-A, 01-1-4-A, OgA-B, 3b1a-C and F6-C.	
Probable members	*G. graminis* virus 45/9-A *Phialophora* sp. with lobed hyphopodia *(Phialophora radicicola* var. *radicicola)* virus 2-2-A	
None (possible genus)	*Gaeumannomyces graminis* virus group II	–
None (possible type species)	*Gaeumannomyces graminis* virus TI-A	–
Main characteristics	A. Properties of the Virus Particle Nucleic acid: Two pieces of linear dsRNA, MW $1.39–1.60 \times 10^6$, separately encapsidated. Some isolates contain additional dsRNA species. Protein: Single major capsid polypeptide species, MW $68–73 \times 10^3$. RNA polymerase (transcriptase) present. Physicochemical properties: $S_{20w} = 133–140$.	

Taxonomic status	*English vernacular name*	*International name*
	Morphology: Diameter 35 nm. Antigenic properties: Serological relationships between members. Serologically unrelated to *G. graminis* group I viruses. B. Replication The virion-associated RNA polymerase catalyzes in vitro transcription of dsRNA, probably by a semi-conservative displacement mechanism. C. Biological Aspects Transmission: Congenital within conidia, and by hyphal anastomosis. Infrequent transmission in ascospores.	
Other members	*G. graminis* viruses F6-B, OgA-A, 3b1a-B	
Probable member	Mushroom virus 4 (from *Agaricus bisporus*)	

References

Bozarth, R.F.: The physicochemical properties of mycoviruses; in Lemke, Viruses and plasmids in fungi, pp. 43–91 (Dekker, New York 1979).

Buck, K.W.; Almond, M.R.; McFadden, J.J.P.; Romanos, M.A.; Rawlinson, C.J.: Properties of thirteen viruses and virus variants obtained from eight isolates of the wheat take-all fungus, *Gaeumannomyces graminis* var. *tritici*. J. gen. Virol. *53:* 235–245 (1981).

Buck, K.W.; McGinty, R.M.; Rawlinson, C.J.: Two serologically unrelated viruses isolated from a *Phialophora* sp. J. gen. Virol. *55:* 235–239 (1981).

Taxonomic status	English vernacular name	International name
None (possible family)	Isometric dsRNA mycoviruses requiring three RNA segments for replication	–
None (possible genus)	*Penicillium chrysogenum* virus group	–
None (possible type species)	*Penicillium chrysogenum* virus	–
Main characteristics	A. Properties of the Virus Particle Nucleic acid: Three pieces of linear monocistronic dsRNA, each of MW$\simeq 2 \times 10^6$, separately encapsidated and comprising$\simeq$16–18% of particle weight. G+C=51%. Some virus isolates may contain an additional dsRNA component. Protein: Single major capsid structural polypeptide species, MW $11–12 \times 10^4$. RNA polymerase present. Lipid: None. Carbohydrate: None. Physicochemical properties: MW$\simeq 11 \times 10^6$; $S_{20w}=150$. Particles disrupted by sodium dodecyl sulfate. Morphology: Polyhedral, 35–40 nm diameter, single shell of protein. Antigenic properties: Serologically related to closely similar viruses of *P. brevicompactum* and *P. cyaneo-fulvum*. B. Replication Virus particles occur in cytoplasm, often in regular aggregates in vesicles or vacuoles. C. Biological Aspects Transmission: Congenital within conidia, and by hyphal anastomosis.	
Other members	Viruses of *P. brevicompactum* and *P. cyaneo-fulvum*	
Possible member	*Helminthosporium victoriae* virus HvV-B	

References

Bozarth, R.F.: The physicochemical properties of mycoviruses; in Lemke, Viruses and plasmids in fungi, pp. 43–9 (Dekker, New York 1979).

Sanderlin, R.S.; Ghabrial, S.R.: Physicochemical properties of two distinct types of virus-like particles from *Helminthosporium victoriae*. Virology *87:* 142–151 (1978).

Taxonomic status	*English vernacular name*	*International name*
None (possible family)	Bisegmented dsRNA animal virus group	–
Main characteristics (from the proposed type species)	A. Properties of the Virus Particle Nucleic acid: Two pieces of linear dsRNA with MWs = 2.5×10^6 and 2.3×10^6 in a single particle. Base composition: larger – 26.7 C, 21.9 A, 28.1 G, 23.3 U; smaller – 27.0 C, 22.8 A, 26.4 G, 23.8 U. RNA is 8.7% by weight of the virus and is not infectious. Protein: Four polypeptides with MWs = 105, 54, 31, and 29×10^3. Number of protein subunits in the virus particle are: VP105 = 22; VP54 = 544; VP31 = 550; VP29 = 122. A dsRNA-dependent RNA transcriptase is present in all three viruses. Protein is 91.3% by weight of virus. Lipid: None present. Carbohydrate: Not known. Physicochemical properties: MW of virus = 55×10^6; $S_{20w} = 435$. Effective buoyant density in CsCl = 1.32 g/cm^3. Stable between pH 3 and 9. Infectivity resistant to 1% SDS at 20° and pH 7.5 for 30 min. Morphology: 60 nm diameter icosahedral particles with 92 morphological subunits. No envelope, no surface projections. Cores are 45 nm in diameter as seen in thin section of infected cells. Cores cannot be generated by treating purified virus with EDTA, trypsin or chymotrypsin. Antigenic properties: A number of distinct antigenic molecules in the virus; one or more antigens associated with the virus surface. Weak hemagglutination using red blood cells of certain inbred mice. B. Replication Viral proteins accumulate in the cytoplasm. Two nonstructural proteins are found: MWs = 63×10^3 (the precursor of VP54) and 29×10^3 (which is qualitatively different from VP29). Virus-induced ssRNA polymerase and dsRNA polymerase appear. Protease(s) cleave the 63×10^3 MW precursor to give VP54 during virus maturation. They cleave some of VP31 to produce VP29, and degrade the 29×10^3 MW nonstructural protein intracellularly. Transcription of viral RNA involves the synthesis of two genome-length mRNA species (24S) that lack a polyA 3′ end tail. There is no inhibition of host cell macromolecular synthesis. Virus particles accumulate in the cytoplasm. Cells lyse but about half the virus remains cell-associated. ds segment reassortment has been demonstrated in laboratory experiments. C. Biological Aspects Host range: Different viruses infecting fish (mostly salmonids); molluscs; chickens, ducks and turkeys; and *Drosophila*. Transmission: Both horizontal and vertical for all viruses. No biological or mechanical vectors known.	
None (possible genus)	Infectious pancreatic necrosis virus genus	–
None (possible type species)	Infectious pancreatic necrosis virus of fish	–
Main characteristics	Same as family description.	

Taxonomic status	*English vernacular name*	*International name*
Other members of possible family	Drosophila X Infectious bursal disease virus of chickens	

References

Becht, H.: Infectious bursal disease virus II. Curr. Top. Microbiol. Immunol. *90:* 107–121 (1980).

Dobos, P.; Hill, B.J.; Hallett, R.; Kells, D.T.; Becht, H.; Teninges, D.: Biophysical and biochemical characterization of five animal viruses with bisegmented dsRNA genomes. J. Virol. *32:* 593–605 (1979).

Pilcher, K.S.; Fryer, J.L.: Viral diseases of fish: a review through 1978. CRC Critical reviews in microbiology, vol. 1 (CRC Press, West Palm Beach 1980).

Taxonomic status	*English vernacular name*	*International name*
Family	–	***TOGAVIRIDAE***
Main characteristics	A. Properties of the Virus Particle Nucleic acid: Single molecule of positive-sense ssRNA, MW about 4×10^6; 5–8% by weight of virus. Protein: 3–4 polypeptides, one or more of which are glycosylated. Lipid and carbohydrate: The virus-specific glycopeptide(s) are inserted in the lipoprotein envelope, whose lipids are cell-derived. Physicochemical properties: $S_{20w}=150$–300. Buoyant density in CsCl$\leq$1.25 g/cm^3; in sucrose, 1.13–1.24 g/cm^3. Morphology: Spherical, 40–70 nm in diameter, with an envelope tightly applied to a spherical nucleocapsid 25–35 nm in diameter with proven or presumed icosahedral symmetry. Surface projections are demonstrable in most togaviruses. Antigenic properties: Members of a genus are serologically related to each other but not to other members of the family. Effects on cells: Members of the genera *Alphavirus, Flavivirus,* and *Rubivirus* show ion-dependent hemagglutinating activity. B. Replication Multiply in cytoplasm and either proven or presumed to mature by budding. C. Biological Aspects All species of the genus *Alphavirus* and most species of the genus *Flavivirus* multiply in arthropods as well as in vertebrates. Members of the genera *Rubivirus* and *Pestivirus* and other possible members of the family are not arthropod-borne. A virus resembling ***Togaviridae*** has been isolated from higher plants (carrot mottle virus).	
Genera	Arbovirus group A Arbovirus group B Rubella virus Mucosal disease virus group	*Alphavirus* *Flavivirus* *Rubivirus* *Pestivirus*
Genus	Arbovirus group A	*Alphavirus*
Type species	Sindbis virus	–
Main characteristics	A. Properties of the Virus Particle Nucleic acid: 4.0–4.6 $\times 10^6$; 5.5–6.3% by weight of particle. Protein: One nonglycosylated capsid protein, MW$\simeq$30–34 $\times 10^3$, and two envelope glycoproteins, MW$\simeq 50 \times 10^3$, have been regularly found. These comprise 57–61% by weight of particle. Lipid and carbohydrate: 27–31% and 6.4% by weight, respectively. Located in the viral membrane. Physicochemical properties: MW$\simeq 70 \times 10^6$. $S_{20w}=280$. Density in sucrose = 1.21 g/cm^3. Morphology: Alphavirus particles have an overall diameter of 70 nm. B. Replication Alphaviruses multiply in the cytoplasm and mature by budding of preassembled nucleocapsids through the plasma membrane.	

Taxonomic status	*English vernacular name*	*International name*
Other members	Aura Bebaru Bijou Bridge Cabassou Chikungunya Eastern equine encephalomyelitis Everglades Fort Morgan Getah Kyzylagach Mayaro Middelburg Mucambo Ndumu O'Nyong-nyong Pixuna Ross River Sagiyama Semliki Forest Tonate Una Venezuelan equine encephalomyelitis Western equine encephalomyelitis Whataroa	
Genus	Arbovirus group B	*Flavivirus*
Type species	Yellow fever virus	–
Main characteristics	A. Properties of the Virus Particle Nucleic acid: MW$\simeq$4.0–4.6 $\times 10^6$. Protein: One glycosylated envelope protein (MW = 53–63 $\times 10^3$); the nucleocapsid protei (MW = 1.35 $\times 10^3$); and a small polypeptide (MW = 7.7–8.5 $\times 10^3$). Physicochemical properties: $S_{20w} \simeq 200$. Morphology: 40–50 nm diameter. B. Replication In the cytoplasm. Maturation presumed to occur by budding through intracytoplasmi membranes (mostly endoplasmic reticulum). C. Biological Aspects Host range: Some multiply in mosquitoes, some in ticks. Some have no known arthropo host. Transmission: Transovarial. Some that are arthropod-borne are sometimes transmitte without vector involvement.	
Other members	These can be divided into subgroups: *Mosquito-borne* Alfuy Bagaza	

Taxonomic status	*English vernacular name*	*International name*
	Banzi	
	Bouboui	
	Bussuquara	
	Dengue, types 1, 2, 3 and 4	
	Edgehill	
	Ilheus	
	Japanese encephalitis	
	Jugra	
	Kokobera	
	Kunjin	
	Murray Valley encephalitis	
	Ntaya	
	Sepik	
	Spondweni	
	St. Louis	
	Stratford	
	Tembusu	
	Uganda S	
	Usutu	
	Wesselsbron	
	West Nile	
	Zika	
	Tick-borne	
	Kadam	
	Karshi	
	Kyasanur Forest disease	
	Langat	
	Louping ill	
	Omsk hemorrhagic fever	
	Powassan	
	Royal Farm	
	Saumarez Reef	
	Tick-borne encephalitis (subtypes: European and Far Eastern)	
	Tyuleniy	
	Vector unknown	
	Apoi	
	Batu Cave	
	Bukalasa bat	
	Carey Island	
	Cowbone Ridge	
	Dakar bat	
	Entebbe bat	
	Israel turkey meningoencephalitis	
	Jutiapa	
	Koutango	
	Modoc	
	Montana myotis leukoencephalitis	
	Negishi	

Taxonomic status	*English vernacular name*	*International name*
	Phnom Penh bat Rio Bravo Rocio Saboya Sokoluk	
Genus	Rubella virus	*Rubivirus*
Type species	Rubella virus	–
Main characteristics	Physicochemically typical of ***Togaviridae*** but serologically unrelated to other ***Togaviridae***. No invertebrate host; man is the only known vertebrate host. Congenital transmission occurs. Only the type species recognized so far.	
Genus	Mucosal disease virus group	*Pestivirus*
Type species	Mucosal disease virus (bovine virus diarrhea virus)	–
Main characteristics	Physicochemically typical of ***Togaviridae*** but do not multiply in invertebrates. Serological cross-reactions occur within genus but not with other members of family ***Togaviridae***. Congenital transmission occurs with hog cholera virus.	
Other members	Border disease (closely related to mucosal disease) Hog cholera (European swine fever)	
Other possible members of family	*Aedes albopictus* cell fusing agent (from *A. aegypti* mosquitoes) Carrot mottle Equine arteritis Lactic dehydrogenase Simian hemorrhagic fever	
Derivation of names	toga: Latin *toga*, 'gown, cloak' alpha: Greek letter 'A' flavi: from Latin *flavus*, 'yellow' rubi: from Latin *rubeus*, 'reddish' pesti: from Latin *pestis*, 'plague'	

References

Berge, T.O.: International catalogue of arboviruses including certain other viruses of vertebrates. US Department of Health, Education and Welfare, DHEW Publ. No. (CDC) 75-8301 (1975).

Horzinek, M.C.: The structure of togaviruses. Prog. med. Virol., vol. 16, pp. 109–156 (Karger, Basel 1973a).

Horzinek, M.C.: Comparative aspects of togaviruses. J. gen. Virol. *20:* suppl., pp. 87–103 (1973b).

Horzinek, M.C.: Non-arthropod-borne togaviruses (Academic Press, New York 1981).

Kääriäinen, L.; Söderlund, H.: Structure and replication of α-viruses. Curr. Top. Microbiol. Immunol. *82:* 15–69 (1978).

Mussgay, M.; Enzmann, P.-J.; Horzinek, M.C.; Weiland, E.: Growth cycle of arboviruses in vertebrate and arthropod cells. Prog. med. Virol., vol. 19, pp. 257–323 (Karger, Basel 1975).
Pfefferkorn, E.R.; Shapiro, D.: Reproduction of togaviruses; in Fraenkel-Conrat, Wagner, Comprehensive virology, vol. 2, pp. 171–230 (Plenum, New York 1974).
Schlesinger, R.W. (ed.): The togaviruses (Academic Press, New York 1980).
Theiler, M.; Downs, W.G.: The arthropod-borne viruses of vertebrates (Yale University Press, New Haven 1973).

Taxonomic status	*English vernacular name*	*International name*
Family	Coronavirus group	***CORONAVIRIDAE***

Main characteristics

A. Properties of the Virus Particle

Nucleic acid: One molecule of infectious ssRNA; MW usually between 5.5×10^6 and 6.1×10^6, although some reports for avian infectious bronchitis virus (IBV) are as high as 8.1×10^6. Polyadenylated at the 3′ terminus.

Protein: Probably 4–6 polypeptides. These fall into three main polypeptide classes, comprising surface projection, envelope or matrix, and nucleocapsid polypeptides, respectively. One of $50–60 \times 10^3$ MW is associated with RNA.

Lipid: Present.

Carbohydrate: Present. High MW surface polypeptides are glycosylated.

Physicochemical properties: Particles band at 1.16–1.23 g/cm^3 in sucrose gradients, with a peak at about 1.18 g/cm^3. Disrupted by ether, chloroform, and detergents. Surface projections removed by bromelain and trypsin.

Morphology: Pleomorphic enveloped particles with average diameter ranging from 75 to 160 nm. Club-shaped surface projections with lengths between 12 and 24 nm protrude from envelope. Internal RNP helix seen as helical structure of diameter 11–13 nm or strands of diameter 9 nm. Particles formed by budding into vesicles.

Antigenic properties: At least two antigenic determinants. Antigenic relationships between different members, e.g., rodent strains.

B. Replication

Six major single-stranded polyadenylated RNA species are produced during replication comprising the intracellular form of the genome and 5 subgenomic RNA species. The genome together with the subgenomic RNAs form a nested sequence set with the sequence of each RNA contained within every larger RNA species. Several virus-specific nonstructural polypeptides have been identified in murine hepatitis virus-infected cells.

C. Biological Aspects

Transmission: Biological – none recognized. Mechanical – IBV, contaminated equipment, personnel, airborne, etc.; porcine transmissible gastroenteritis virus, fecal-oral route. No data for other members.

Genus	Coronavirus	*Coronavirus*
Type species	Avian infectious bronchitis virus (IBV)	–
Other members	Human coronavirus Murine hepatitis Porcine hemagglutinating encephalitis Porcine transmissible gastroenteritis	
Probable members	Canine coronavirus Coronavirus enteritis of turkeys (bluecomb disease) Neonatal calf diarrhea coronavirus Rat coronavirus Sialodacryoadenitis of rats	

Taxonomic status	*English vernacular name*	*International name*
Possible members	Feline infectious peritonitis (feline coronavirus) Human enteric coronavirus	

Derivation of name	corona: Latin 'crown', from appearance of surface projections in negatively stained electron micrographs

References

Bond, C.W.; Leibowitz, J.L.; Robb, J.A.: Pathogenic murine coronaviruses. II. Characterization of the virus-specific proteins of murine coronaviruses JHMV and A59V. Virology *94:* 371–384 (1979).

Siddell, S.; Wege, H.; Barthel, A.; Ter Meulen, V.: Coronavirus JHM: intracellular protein synthesis. J. gen. Virol. *53:* 145–155 (1981).

Spaan, W.J.M.; Rottier, P.J.M.; Horzinek, M.C.; van der Zeijst, B.A.M.: Isolation and identification of virus-specific mRNAs in cells infected with mouse hepatitis virus (MHV-A59). Virology *108:* 424–434 (1981).

Stern, D.F.; Kennedy, S.I.T.: Coronavirus multiplication strategy. I. Identification and characterization of virus-specified RNA. J. Virol. *34:* 665–674 (1980).

Stern, D.F.; Kennedy, S.I.T.: Coronavirus multiplication strategy. II. Mapping of avian infectious bronchitis virus intracellular RNA species to the genome. J. Virol. *36:* 440–449 (1980).

Tyrrell, D.A.J.; Alexander, D.J.; Almeida, J.D.; Cunningham, C.H.; Easterday, B.C.; Garwes, D.J.; Hierholzer, J.C.; Kapikian, A.; Macnaughton, M.R.; McIntosh, K.: Coronaviridae: second report. Intervirology *10:* 321–328 (1978).

Taxonomic status	English vernacular name	International name
Family	–	***PARAMYXOVIRIDAE***

Main characteristics

A. Properties of the Virus Particle

Nucleic acid: One molecule of ssRNA, MW = $5–7 \times 10^6$. About 0.5% by weight of virus particle. Most particles contain a negative-sense strand, but some contain positive-sense strands. Thus, partial self-annealing of isolated RNA may be demonstrated.

Protein: 5–7 polypeptides; MWs $35–200 \times 10^3$. Enzymes (variously represented and reported among genera): transcriptase, polyadenylate transferase, mRNA methyl transferase, neuraminidase. Other functional proteins: hemagglutinin (and neuraminidase, if present), each represented by a single glycoprotein species; cell fusion and hemolysis mediated by another glycoprotein species.

Lipid: 20–25% by weight; composition mainly host cell dependent.

Carbohydrate: 6% by weight; composition host cell dependent.

Physicochemical properties: MW of virion at least 500×10^6, much more for pleomorphic multiploid virions; density in sucrose = 1.18–1.20 g/cm^3; S_{20w} at least 1,000; sensitive to lipid solvents, nonionic detergents, formaldehyde, and oxidizing agents.

Morphology: Shape of virions pleomorphic, but usually roughly spherical, 150 nm or more in diameter; envelope derived from cell surface membrane lipids, incorporating virus glycoproteins and a nonglycosylated virus protein; surface projections, 8 nm long, spaced 8–10 nm apart, contain virus glycoproteins. Nucleocapsid has helical symmetry; diameter 12–17 nm, depending on genus; length 1 μm in some genera.

Antigenic properties: One or more surface antigens, involved in virus neutralization; one nucleocapsid antigen described; specificities of antigens vary among genera.

Effects on cells: Generally cytolytic, but temperate and persistent infections are common; other features are inclusions, syncytium formation, and hemadsorption.

B. Replication

Virus entry by fusion of envelope with cell surface membrane; nucleocapsid is the functional template for transcription of complementary viral mRNA species and for RNA replication; independently assembled nucleocapsids are enveloped at the cell surface at sites containing virus envelope proteins. Virions released by budding.

C. Biological Aspects

Host range: Each virus has its own host range, in nature and in the laboratory.

Transmission: Horizontal, mainly airborne; no vectors.

Genera		
	Paramyxovirus group	*Paramyxovirus*
	Measles-rinderpest-distemper (MRD) group	*Morbillivirus*
	Respiratory syncytial virus group	*Pneumovirus*

Genus	Paramyxovirus group	*Paramyxovirus*
Type species	Newcastle disease virus	–

Main characteristics: All members of the genus possess a neuraminidase, in contrast to members of the other two genera.

Other members		
	Finch paramyxovirus	avian
	Mumps	human
	Nariva	murine

Taxonomic status	English vernacular name		International name
	Parainfluenza 1 (Sendai)	human, murine	
	Parainfluenza 2	human	
	Parainfluenza 3	human, bovine, ovine	
	Parainfluenza 4	human	
	Parainfluenza 5	avian, simian, canine	
	Parainfluenza turkey/Ontario	avian	
	Parrot paramyxovirus	avian	
	Yucaipa	avian	
Genus	Measles-rinderpest-distemper (MRD) group		*Morbillivirus*
Type species	Measles virus		–
Main characteristics	All members lack neuraminidase, which genus *Paramyxovirus* possesses, and differ from genus *Pneumovirus* in size of nucleocapsid and other structural features. All members produce both cytoplasmic and intranuclear inclusion bodies which contain viral RNP. Members of the group are clearly related antigenically.		
Other members	Canine distemper	canine	
	Peste-des-petits-ruminants (PPR)	ovine	
	Rinderpest	bovine	
Genus	Respiratory syncytial virus group		*Pneumovirus*
Type species	Respiratory syncytial virus		–
Main characteristics	Lacks neuraminidase. Differs from *Paramyxovirus* and *Morbillivirus* genera in size of nucleocapsid and other structural features.		
Other members	Bovine respiratory syncytial		
	Pneumonia of mice		
Derivation of names	paramyxo:	from Greek *para,* 'by the side of', and *myxa,* 'mucus' (relating to activity of hemagglutinin and neuraminidase)	
	morbilli:	plural of Latin *morbillus,* diminutive of *morbus,* 'disease', hence 'measles'	
	pneumo:	from Greek *pneuma,* 'breath'	

References

Appel, M.J.G.; Gillespie, J.H.: Canine distemper virus. Virology Monographs, vol. 11 (Springer, New York 1972).

Choppin, P.W.; Compans, R.W.: Reproduction of paramyxoviruses; in Fraenkel-Conrat, Wagner, Comprehensive virology, vol. 4, pp. 95–178 (Plenum, New York 1975).

Compans, R.W.; Harter, D.H.; Choppin, P.W.: Studies on pneumonia virus of mice (PVM) in cell culture. II. Structure and morphogenesis of the virus particle. J. exp. Med. *126:* 267–276 (1967).

Hsiung, G.D.: Parainfluenza-5 virus. Infection of man and animal. Prog. med. Virol., vol. 14, pp. 241–274 (Karger, Basel 1972).

Jackson, G.G.; Muldoon, R.L.: Viruses causing common respiratory infections in man. II. Enteroviruses and paramyxoviruses. J. infect. Dis. *128:* 387–469 (1973).

Mahy, B.W.J.; Barry, R.D. (eds.): Negative strand viruses and the host cell (Academic Press, London 1978).

Taxonomic status	*English vernacular name*	*International name*
Family	Influenza virus group	***ORTHOMYXOVIRIDAE***
Genus	Influenza virus	*Influenzavirus*
Type species	Influenza virus A/WS/33 (H0N1)	–

Main characteristics

A. Properties of the Virus Particle

Nucleic acid: Eight molecules of linear negative-sense ssRNA; MWs$\simeq$0.2–1 × 10^6 fo individual segments. Total MW$\simeq$5 × 10^6 for influenza A and B viruses; proportion o individual segments in virus preparation may vary according to ratio of complete: incom plete particles; additional RNA species, including host-derived RNA and possible frag ments of larger virion RNA segments, may also be present in some preparations. Percen base composition for influenza A or B virus: C, 23–26; A, 21–23; G, 18–20; U, 31–36.

Protein: 7–9 polypeptides in virions depending on extent of proteolytic cleavage of hemag glutinin (HA) polypeptide (MW$\simeq$60,000, plus carbohydrate) into HA1 (MW$\simeq$35,00 plus 25% carbohydrate) and HA2 (MW$\simeq$25,000 plus 10% carbohydrate) component which are covalently linked to each other by disulfide bonds. Enzymic activities are RNA dependent RNA polymerase (internal) (three polypeptides P1, P2, P3; MWs$\simeq$80–100 × 10^3 and receptor-destroying enzyme [external neuraminidase (NA) in influenza A and I viruses]. Functional NA is tetramer of NA polypeptide (MW$\simeq$50,000 plus carbohydrate) External hemagglutinating activity caused by proteins which are reportedly trimers of HA and/or HA1 plus HA2 polypeptides. Virus infectivity is enhanced by specific proteolyti cleavage of HA into HA1 plus HA2. Numbers of proteins in virions variable. NP, o nucleoprotein, MW$\simeq$50,000. M, 'matrix' or 'membrane' protein, MW$\simeq$25,000. Non structural protein(s) (NS) in infected cells (MW$\simeq$20,000).

Lipid: 18–37% by weight of virion; composition host-dependent but minus sialic acid du to the action of virus NA. Present in virion envelope.

Carbohydrate: 5–9% by weight of virion, composition host- and virion-dependent (minu sialic acid). Present on external polypeptides (HA and NA) which may also contain coval ently bound sulfate.

Physicochemical properties: MW = 250 × 10^6; S_{20w} of nonfilamentous particles$\simeq$700–800 density in sucrose/H_2O = 1.19 g/cm^3. Virus infectivity reduced within minutes by exposur to low pH (<5) or heat (56°). Lipid solvents and detergents (anionic, cationic or neutral destroy membrane integrity with resultant reduction in infectivity. Infectivity may be totall destroyed by treatment with formaldehyde, β-propiolactone, UV light and gamma irradia tion, without affecting antigenic specificity. Prolonged exposure to chemicals and radiatio inactivates different replicative events at different rates, presumably as a result of productio of lesions in individual RNA segments of different sizes.

Morphology: Nucleocapsid of helical symmetry and diameter 9–15 nm enclosed withi lipoprotein membrane having surface projections. Nucleoproteins of different size classe (50–130 nm length), with loop at each end, extractable from virions or infected cells Arrangement within virion uncertain, although coils of about 4–20 turns of a 7 nm thic material are sometimes seen in partially disrupted virus. Virions are pleomorphic, abou 80–120 nm in diameter, but the length is up to several micrometers ('filamentous form') M protein believed to form layer on inside of lipid layer, with HA and NA glycoprotein projecting about 8 nm from surface. Isolated HA protein rods of 14 nm length and 4 nm diameter. Isolated NA protein, when intact, comprises head of dimension 4 × 8.5 nm attached to stalk 10 × 4 nm. Cores containing M, RNP and P proteins may be generate by controlled chemical disruption of virions.

Taxonomic status	*English vernacular name*	*International name*

Antigenic properties: Known antigens include NP, M, HA, NA, and NS. NP and M are type-specific for influenza A and B strains. Variation occurs within HA and NA antigens. Ten or eleven subgroups of HA and eight subgroups of NA recognized for influenza A viruses, with minimal serological cross-reaction between subgroups. Additional variation occurs within subgroups, particularly for viruses isolated in different years, although only a small number of strains of any subgroup are present at any time. HA and NA antigens of influenza B viruses exhibit less antigenic variation than for influenza A, and no subgroups are defined. Antibody to HA neutralizes infectivity. Antibody to NA has poor neutralizing activity; if present during multicycle replication, may give apparent neutralization by reducing virus yield from infected cells.
Effects of virus suspensions on cells: Erythrocytes of many species agglutinated by virions. Sialic acid-containing virus receptors on erythrocytes may be destroyed by NA of attached virions, resulting in elution of virus.

B. Replication
1. Virions attach to sialic acid-containing receptors on plasma membrane of host cell and enter by fusion of virion envelope with plasma membrane, or by uptake in vacuoles formed from cell plasma membrane. Transcriptase complex synthesizes complementary RNA transcripts of infecting virus genome, requiring prior DNA-dependent RNA synthesis by host cell that may be inhibited by actinomycin D or α-amanitin. Complementary RNA functions as template for viral-specific protein synthesis, and also as template for replication of virion strand RNA by viral-specific replicase. Nucleoprotein antigens accumulate in nucleus during first hours of infection, then migrate to cytoplasm. HA and NA polypeptides associate with cytoplasmic membranes, where glycosylation occurs. Site of cleavage of HA polypeptide to HA1 and HA2 is unresolved. HA and NA proteins migrate to localized regions of plasma membrane, where new virions form by budding, incorporating M protein on the inner surface of the envelope and usually incorporating RNPs. Reassortment of genes between viruses of the same type readily occurs in mixed infections, producing new genetic combinations.
2. Inclusion bodies identified in some virus-cell systems contain crystalline NS protein.

C. Biological Aspects
Host range: Cause epidemic or sporadic influenza in man. Certain influenza virus A strains cause epizootics of respiratory infections in pigs and horses. Other influenza virus A strains cause enzootic and epizootic generalized or respiratory infections of birds. Dogs, cattle, bears, primates and seals reported to be infected sporadically with influenza virus A. Natural interspecies transfer described from man to pigs and less frequently from pigs to man. Influenza B strains infect primarily man. Some evidence of occasional infection of domestic animals in close contact with man. Ferrets, mice, hamsters and guinea pigs may be artificially infected. Most strains grow on primary monkey, human, chick, and calf kidney cells. Certain continuous cell lines may also be used with addition of trypsin. Host specificity for nonhuman strains not predictable.
Transmission: Aerosol, waterborne, and direct contact.

Other member	Influenza B/Lee/40	
Probable genus	Influenza virus C	–
Main characteristics	ssRNA; 4–6 segments. Total RNA MW$=4.15–5.15 \times 10^6$. Percent base composition:	

C, 20; A, 23; G, 20; U, 36. At least five polypeptides in virion, three glycosylated. Receptor-destroying enzyme not neuraminidase. Physicochemical and morphological characteristics similar to A and B viruses, but not proven that hemagglutinin and receptor-destroying enzyme located on separate structures.

Derivation of names	ortho:	from Greek *orthos,* 'straight, correct'
	myxo:	from Greek *myxa,* 'mucus'
	influenza:	Italian form of Latin *influentia,* 'epidemic', so used because epidemics were thought to be due to astrological or other occult 'influence'

References

Compans, R.W.; Bishop, D.H.L.; Meier-Ewert, H.: Structural components of influenza C virions. J. Virol. *21:* 658–665 (1977).

Francis, T., Jr.: A new type of virus from epidemic influenza. Science *92:* 405–408 (1940).

Francis, T., Jr.; Quilligan, J.J.; Minuse, E.: Identification of another epidemic respiratory disease. Science *112:* 495–497 (1950).

Horsfall, F.L.; Lennette, E.H.; Rickard, E.R.; Andrewes, C.H.; Smith, W.; Stuart-Harris, C.H.: The nomenclature of influenza. Lancet *ii:* 413–414 (1940).

Hoyle, L.: The influenza viruses. Virology Monographs, vol. 4 (Springer, New York 1968).

Kendal, A.P.: A comparison of 'influenza C' with prototype myxoviruses: receptor-destroying activity (neuraminidase) and structural polypeptides. Virology *65:* 87–99 (1975).

Kilbourne, E.D. (ed.): The influenza viruses and influenza (Academic Press, New York 1975).

Ritchey, M.B.; Palese, P.; Kilbourne, E.D.: RNAs of influenza A, B and C viruses. J. Virol. *18:* 738–744 (1976).

Scholtissek, C.: The genome of the influenza virus. Curr. Top. Microbiol. Immunol. *80:* 139–169 (1978).

Skehel, J.J.; Hay, A.J.: Influenza virus transcription. J. gen. Virol. *39:* 1–8 (1978).

Stuart-Harris, C.H.; Schild, G.C.: Influenza: the virus and the disease (Arnold, London 1976).

Taylor, R.M.: Studies on survival of influenza virus between epidemics and antigenic variants of the virus. Am. J. publ. Hlth. *39:* 171–178 (1949).

WHO Study Group: A revised system of nomenclature for influenza viruses. Bull. Wld Hlth Org. *45:* 119–124 (1971).

Taxonomic status	*English vernacular name*	*International name*
Family	Bullet-shaped virus group	***RHABDOVIRIDAE***

Main characteristics A. Properties of the Virus Particle

Nucleic acid: One molecule of noninfectious linear (negative-sense) ssRNA. $S_{20w}=38–45$. MW=$3.5–4.6\times10^6$; 1–2% by weight of virus.

Protein: 4–5 major polypeptides [designated L, G, N, NS and M for vesicular stomatitis virus (VSV)]; 65–75% by weight of the virus. Other polypeptides may be present in minor amounts. Transcriptase and other enzymes present in virus.

Lipid: 15–25% by weight of virus, the composition being dependent on the host cell.

Carbohydrate: 3% by weight of virion. Associated with surface projections and glycolipids; minor variation with host cell type.

Physicochemical properties: MW$\simeq$300–1,000$\times10^6$. $S_{20w}\simeq$550–1,000. Density in CsCl= 1.19–1.20 g/cm^3; in sucrose, 1.17–1.19 g/cm^3. Infcetivity: unstable at pH 3; stable in the range pH 5–10. Rapidly inactivated at 56° and by UV- and X-irradiation. Sensitive to lipid solvents.

Morphology: The viruses infecting vertebrates and invertebrates are usually bullet-shaped, and those infecting plants are usually bacilliform. Dimensions: 130–380 nm$\times$50–95 nm, with surface projections (G protein) 5–10 nm long and$\simeq$3 nm in diameter. In thin section, a central axial channel is seen. Characteristic cross-striations (spacing 4.5 to 5.0 nm) seen in negatively stained and thin-sectioned particles. Truncated particles 0.1–0.5 of the length of the virus may be common except in members infecting plants. Abnormally long and double-length particles and tandem formations are sometimes observed. A honeycomb pattern is observed on the surface of some members. The inner nucleocapsid, about 50 nm in diameter, with helical symmetry, consists of an RNA-N protein complex together with L and NS proteins, surrounded by M protein. The nucleocapsid contains a transcriptase and is infectious. It uncoils to a helical structure 20$\times$700 nm.

Antigenic properties: The glycoprotein is involved in virus neutralization. G protein defines serotype. N protein shows cross-reactions in VSV and in lyssaviruses. N antigen apparently different in two serotypes of potato yellow dwarf virus.

B. Replication

Viral proteins accumulate in the cytoplasm except for some plant members. Virus RNA is transcribed by virion transcriptase into several positive-strand RNA species which act as mRNA in polyribosome complexes. Virus RNA replication details not known, but may involve RNA-N template intermediate. Nucleocapsid synthesis and envelope then formed by insertion of virus protein G into preexisting host cell membrane. Site of formation of mature particles variable, depending on virus and host cell – e.g., VSV nucleocapsid is synthesized in cytoplasm and then predominantly buds from the plasma membrane in most, but not all, cells; rabies predominantly from intracytoplasmic membranes; and potato yellow dwarf virus from the inner nuclear membrane. Complete particles of this virus accumulate in the perinuclear space.

C. Biological Aspects

Host range: Some members multiply in arthropods as well as vertebrates, others multiply in arthropods and plants. Biological and/or mechanical vectors include mosquitoes, sandflies, tabanids, midges, mites, leafhoppers and aphids. Sigma virus recognized first as a congenital infection of Drosophila. Experimental: in vivo – some vertebrate members have a very wide host range; in vitro – wide range of vertebrate and invertebrate cells susceptible

Taxonomic status	*English vernacular name*	*International name*
	to vertebrate viruses. Potato yellow dwarf virus multiplies in leafhopper cells. Plant members usually have narrow host range among higher plants. Transmission: Some viruses transmitted vertically in insects, but none is so transmitted in vertebrates or plants. Some mechanically transmitted in plants. Vector transmission by mosquitoes, sandflies, culicoides, mites, aphids, or leafhoppers. Mechanical transmission by contact or aerosol, bite, or venereal.	
Genus	Vesicular stomatitis virus group	*Vesiculovirus*
Genus	Rabies virus group	*Lyssavirus*
None	Plant rhabdoviruses	–

Genus	Vesicular stomatitis virus group	*Vesiculovirus*
Type species	Vesicular stomatitis (Indiana serotype)	–

Main characteristics

A. Properties of the Virus Particle

Nucleic acid: The genome of VSV-Indiana consists of 5 genes in tandem with no overlaps in the order (5′) L-G-M-NS-N (3′). All but 70 of the approximately 10,000 nucleotides are represented in plus-strand transcripts comprising five monocistronic mRNAs plus an untranslated leader sequence of 48 nucleotides. The untranscribed regions are a 59 nucleotide (3′)-terminal region of the L gene, a 3 nucleotide spacer between leader and N gene, and 4 dinucleotide spacers (CA or GA) at the four inter-cistronic junctions. There is a common unadecamer sequence (5′) UAUGAAAAAAA (3′) preceding each intercistronic junction, and the sequences complementary to the (5′)-end of each mRNA have the general form (5′) AACAGNNAUC (3′).

Proteins: L (large) MW$\simeq 150 \times 10^3$; G (glycoprotein) MW$\simeq 70$–80×10^3; N (in nucleoprotein) MW$\simeq 50$–62×10^3; NS (nonstructural) MW$\simeq 40$–50×10^3; M (matrix) MW$\simeq 20$–30×10^3. Number of protein subunits in virion: L, 20–50; G, 500–1,500; N, 1,000–2,000; NS, 100–300; M, 1,600–4,000. Enzymes in virion: transcriptase (made up of L plus NS proteins); protein kinase (host?); guanyl and methyl transferase; nucleotide triphosphatase; nucleoside diphosphate kinase; 5′ capping enzyme.

Physicochemical properties: $S_{20w} = 625$. Virus is 170 nm long, 70 nm wide.

Morphology: Helix of the nucleocapsid has an outer diameter of 49 nm; inner diameter = 29 nm; 35 subunits per turn. The RNP is linear and 1 μm long.

Antigenic properties: G protein functions as type-specific immunizing antigen; N is a cross-reacting CF antigen.

B. Replication

VSV replicates in enucleate cells. Phenotypic mixing is extensive between VSV and heterologous lytic viruses (simian virus 5, Newcastle disease virus, fowl plague virus, herpes simplex virus), nonlytic viruses (avian myeloblastosis virus, murine leukemia virus, mouse mammary tumor virus), and partially expressed endogenous viruses. Phenotypic mixing (complementation) also occurs within but not between serological types of VSV. Complementation also reported to occur by re-utilization of structural components of UV-irradiated VSV. Complementation shown with VSV (Indiana, Cocal, New Jersey) and Chandipura. Five or six nonoverlapping groups (VSV, Chandipura). Inter-strain complementation only observed with serologically related viruses (VSV-Indiana and Cocal).

Taxonomic status	*English vernacular name*	*International name*
	C. Biological Aspects Indiana serotype isolated from vertebrates and insects.	
Other members	Isolated in nature from vertebrates (V) or from invertebrates (I): Vesicular stomatitis (further subtypes and serotypes) – Argentina (V) – Brazil (Alagoas) (V) – Cocal (V, I) – New Jersey (V, I) Chandipura (V, I) Isfahan (V, I) Piry (V)	
Genus	Rabies virus group	*Lyssavirus*
Type species	Rabies virus	–
Main characteristics	Morphologically similar to *Vesiculovirus* but antigenically distinct. Contains 5 proteins: L, MW$\simeq 190 \times 10^3$; G, MW$\simeq 65–80 \times 10^3$; N, MW$\simeq 58–62 \times 10^3$; M1 (=NS), MW$\simeq 35–40 \times 10^3$; M2, MW$\simeq 22–25 \times 10^3$. Number of protein subunits in virion: L, 17–150 (strain-dependent); G, 1,600–1,900; N, 1,750; M1, 900–950; M2, 1,650–1,700. Enzymes in virion: transcriptase (L+NS). Multiply in vertebrates and insects.	
Other members	Isolated in nature from vertebrates (V) or invertebrates (I): Duvenhage (V) Kotonkan (I) Lagos bat (V) Mokola (V) Obodhiang (I)	
Probable members of the family, but no genera established	Bahia Grande (TB4 1054) (I) Barur (V, I) BeAn 157575 (V) Be Ar 185559 (I) BFN 3187 (Grey Lodge) (I) Bovine ephemeral fever (V, I) Chaco (V) Egtved (viral hemorrhagic septicemia) (V) Flanders (V, I) Hart Park (V, I) Infectious hematopoietic necrosis virus (V) Joinjakaka (I) Kamese (I) Kern Canyon (V) Keuraliba (V) Kimberley (I) Klamath (V) Kununurra (I)	

Taxonomic status	English vernacular name	International name
	Kwatta (I) Marco (V) Mossuril (V, I) Mount Elgon bat (V) Navarro (V) New Minto (I) Oita 293 (V) Parry Creek (I) Red disease of pike (pike fry rhabdovirus) (V) Rhabdovirus of blue crab (I) Rhabdovirus of eels (V) Rhabdovirus of entamoeba (I) Rhabdovirus of grass carp (V) Sawgrass (I) Sigma (I) Spring viremia of carp (V) S-1643 (I) Timbo (V) Yata (I)	
None	Plant rhabdoviruses (244)	–
Main characteristics	Particles are bacilliform and/or bullet-shaped with a distinct prevalence of the bacilliform. Mature virions are 135–380 nm long and 45–95 nm wide. The nucleocapsid is formed by a helically wound nucleoprotein (negative-sense ssRNA plus N protein). Two subgroups (A and B) can be recognized based on site of particle assembly and protein composition. Viruses of subgroup A have properties comparable with *Vesiculovirus*, i.e., they mature and accumulate in the cytoplasm, have protein M ($MW \simeq 18–25 \times 10^3$) and transcriptase activity which is readily detectable in vitro. Protein L ($MW \simeq 145–170 \times 10^3$) is detected in some members of subgroup A. Viruses of subgroup B share some properties with *Lyssavirus*, i.e., they bud at the inner membrane of the nuclear envelope and accumulate perinuclearly, possess protein M1 ($MW \simeq 27–44 \times 10^3$) and M2 ($MW \simeq 21–39 \times 10^3$). Viruses of both groups have protein G ($MW \simeq 71–93 \times 10^3$) and N ($MW \simeq 55–60 \times 10^3$). A third grouping (listed under possible members) is made up of viruses with nonenveloped particles $\simeq$35 nm wide and 100–120 nm long which are usually associated with the nucleus, giving rise to 'spokewheel-like' structures. These particles resemble true rhabdovirus nucleocapsids, but the relationship with rhabdoviruses is yet to be established. Some plant rhabdoviruses are sap-transmissible, all are readily inactivated at room temperature. Multiplication occurs in aphid or leafhopper vectors as well as plants.	
Members	Characterized physicochemically and transmitted experimentally. Vector indicated where known: *Subgroup A* – Lettuce necrotic yellows (aphid) (type species) (26) Broccoli necrotic yellows (aphid) (85) *Sonchus* Wheat striate mosaic (hopper) (99)	

Taxonomic status	*English vernacular name*	*International name*
	Subgroup B – Potato yellow dwarf (hopper) (type species) (35)	
	Eggplant mottled dwarf (115)	
	Sonchus yellow net (aphid) (205)	
	Sowthistle yellow vein (aphid) (62)	
Probable members officially ungrouped but listed according to type of vector (where known)	Transmitted experimentally but not characterized physicochemically. Vector indicated where known:	
	1. Aphid – Carrot latent	
	Lucerne enation	
	Parsley latent	
	Raspberry vein chlorosis (174)	
	Strawberry crinkle (163)	
	2. Hopper – Barley yellow striate mosaic (cereal striate)	
	Cereal chlorotic mottle	
	Colocasia bobone disease	
	Digitaria striate	
	Finger millet mosaic	
	Maize mosaic (94)	
	Northern cereal mosaic	
	Oat striate	
	Rice transitory yellowing (100)	
	Russian winter wheat mosaic	
	Sorghum stunt mosaic	
	Wheat chlorotic streak	
	3. Lace bug – Beet leaf curl	
	4. Mite – Coffee ringspot	
	5. Not known – *Chrysanthemum frutescens*	
	Cow parsnip mosaic	
	Cynara	
	Endive	
	Gomphrena	
	Melilotus latent	
	Pelargonium vein clearing	
	Pisum	
	Pittosporum vein yellowing	
	Raphanus	
Possible members	Recognized only as virus-like particles:	
	Atropa belladonna	
	Callistephus chinensis chlorosis	
	Carnation bacilliform	
	Cassava symptomless	
	Chondrilla juncea stunting	
	Chrysanthemum vein chlorosis	
	Clover enation	
	Euonymus fasciation	
	Festuca leaf streak	
	Gerbera symptomless	

Taxonomic status	*English vernacular name*	*International name*
	Holcus lanatus yellowing	
	Iris germanica leaf stripe	
	Ivy vein clearing	
	Laburnum yellow vein	
	Laelia red leafspot	
	Launea arborescens stunt	
	Lemon scented thyme leaf chlorosis	
	Lolium (ryegrass)	
	Lotus stem necrosis	
	Lupin yellow vein	
	Malva silvestris	
	Melon variegation	
	Patchouli *(Pogostemon patchouli)* mottle	
	Pigeon pea *(Cajanus cajan)* proliferation	
	Pineapple chlorotic leaf streak	
	Plantain *(Plantago lanceolata)* mottle	
	Ranunculus repens symptomless	
	Red clover mosaic	
	Saintpaulia leaf necrosis	
	Sambucus vein clearing	
	Sarracenia purpurea	
	Triticum aestivum chlorotic spot	
	Vigna sinensis mosaic	
	Zea mays	
	Recognized as nonenveloped particles:	
	Citrus leprosis	
	Dendrobium leaf streak	
	Orchid fleck	
	Phalaenopsis chlorotic spot	

Derivation of names		
	rhabdo:	from Greek *rhabdos,* 'rod'
	vesiculo:	from Latin *vesicula,* diminutive of *vesica,* 'bladder, blister'
	lyssa:	Greek 'rage, rabies'

References

Brown, F.; Bishop, D.H.L.; Crick, J.; Francki, R.I.B.; Holland, J.J.; Hull, R.; Johnson, K.; Martelli, G.; Murphy, F.A.; Obijeski, J.F.; Peters, D.; Pringle, C.R.; Reichmann, M.E.; Schneider, L.G.; Shope, R.E.; Simpson, D.I.H.; Summers, D.F.; Wagner, R.R.: Rhabdoviridae. Intervirology *12:* 1–7 (1979).

Dale, J.L.; Peters, D.: Protein composition of the virions of five plant rhabdoviruses. Intervirology *16:* 86–94 (1981).

Francki, R.I.B.; Kitajima, E.W.; Peters, D.: Rhabdoviruses of plants; in Kurstak, Handbook of plant virus infections and comparative diagnosis, pp. 455–489 (Elsevier/North Holland, Amsterdam 1981).

Francki, R.I.B.; Randles, J.W.: Rhabdoviruses infecting plants; in Bishop, Rhabdoviruses, vol. 3, pp. 135–165 (CRC Press, West Palm Beach 1979).

Martelli, G.P.; Russo, M.: Rhabdoviruses of plants; in Maramorosch, The atlas of insect and plant viruses pp. 181–213 (Academic Press, New York 1977).

Taxonomic status	*English vernacular name*	*International name*
Family	–	***BUNYAVIRIDAE***

Main characteristics

A. Properties of the Virus Particle

Nucleic acid: Three molecules of negative-sense ssRNA (large, medium, small: L, M, S). Ends are hydrogen-bonded, making molecules circular. MWs about 3–5, 1–2 and 0.4–0.8 $\times 10^6$; 1–2% by weight.

Protein: 3 major, 1 minor, consisting of 2 external glycoproteins (G1, G2), a nucleocapsid protein (N), and minor large protein (L). Transcriptase present.

Lipid: 20–30% by weight; forms envelope.

Carbohydrate: 7% by weight; components of the glycoproteins and glycolipids.

Physicochemical properties: MW = $300–400 \times 10^6$; S_{20w} = 350–470. Density in CsCl = 1.20 g/cm^3. Sensitive to lipid solvents and detergents.

Morphology: Spherical, oval, enveloped particles (90–100 nm diameter) with glycoprotein surface projections, ribonucleocapsids (three) composed of circular, long helical strands, 2–2.5 nm diameter, sometimes supercoiled, lengths of 0.2–3 μm depending on arrangement.

Antigenic properties: CF, hemagglutinin and neutralizing antigenic determinants present on viral glycoproteins.

Effects of virus suspensions on cells: Some induce cell fusion.

B. Replication

Molecular biology: Capable of primary mRNA transcription in absence of protein synthesis. Replicate in cytoplasm. Mature by budding into smooth surfaced vesicles in the Golgi region or nearby. Genetic recombination has been demonstrated for members of Bunyamwera and California serogroups, involving segment reassortment.

C. Biological Aspects

Host range: Various; warm and cold-blooded vertebrates and arthropods.

Transmission: Vectors involving mosquitoes, ticks, *Phlebotomus* and other arthropods. Transovarial and venereal transmission demonstrated for some mosquito-borne viruses. Aerosol infection occurs in selected situations. In some instances, avian host and/or vector movements may result in virus dissemination.

Genera	Bunyamwera supergroup	*Bunyavirus*
	Sandfly fever group	*Phlebovirus*
	Nairobi sheep disease	*Nairovirus*
	Uukuniemi	*Uukuvirus*

Genus	Bunyamwera supergroup	*Bunyavirus*
Type species	Bunyamwera virus	–

Main characteristics

A. Properties of the Virus Particle

Nucleic acid: L RNA, $2.7–3.1 \times 10^6$; M RNA, $1.8–2.3 \times 10^6$; S RNA, $0.28–0.50 \times 10^6$.

Protein: G1, $108–120 \times 10^3$; G2, $29–41 \times 10^3$; N, $19–25 \times 10^3$; L, $145–200 \times 10^3$. Both glycoproteins coded by M RNA; N coded by S RNA. L protein probably coded by L RNA.

B. Replication

At least one nonstructural S RNA gene product synthesized in infected cells.

Taxonomic status	English vernacular name	International name
	C. Biological Aspects Host range: Various vertebrate species; primarily mosquito, but occasional alternate arthropod species, *Culicoides,* phlebotomines, ticks and horseflies. Virulence: Virulence capability primarily determined by viral M RNA gene products (glycoproteins); virulence can be mitigated by defects in L RNA gene products.	
Other members (probable members indicated by asterisk)	There are 16 serological groups of the genus *Bunyavirus* (at least 145 viruses) and some ungrouped viruses serologically unrelated to members of other genera. Mostly mosquito-transmitted; some (Tete group) tick-transmitted. Some capable of transovarial transmission. The groups are: Anopheles A group: (11) Anopheles A, Lukuni, Tacaiuma, Virgin River, CoAr3624*, CoAr1071*, CoAr3627*, ColAn57389*, SpAr2317*, H32580*, AG80–24* Anopheles B group: (2) Anopheles B, Boraceia Bunyamwera group: (23) Anhembi, Batai, Birao, Bunyamwera, Cache Valley, Calovo, Germiston, Guaroa, Ilesha, Kairi, Lokern, Maguari, Main Drain, Northway, Playas*, Santa Rosa, Shokwe, Sororoca, Taiassui, Tensaw, Tlacotalpan, Wyeomyia, CbaAr426* Bwamba group: (2) Bwamba, Pongola C group: (14) Apeu, Bruconha*, Caraparu, Gumbo Limbo, Itaqui, Madrid, Marituba, Murutucu, Nepuyo, Oriboca, Ossa, Restan, Vinces*, 63U11* California group: (14) California encephalitis, Inkoo, Jamestown Canyon, Jerry Slough, Keystone, La Crosse, Lumbo*, Melao, San Angelo, Serra do Navio, snowshoe hare, South River*, Tahyna, Trivittatus Capim group: (9) Acara, Benevides, Benfica, Bushbush, Capim, Guajara, Juan Diaz, Moriche, GU71u344* Gamboa group: (7) Alajuela*, Gamboa, Pueblo Viejo*, San Juan*, 75V–2621*, 78V–2441*, 75V–2374* Guama group: (12) Ananindeua, Bertioga, Bimiti, Cananeia*, Catu, Guama, Guaratuba, Itimirim, Mahogany Hammock, Mirim, Moju, Timboteua Koongol group: (2) Koongol, Wongal Minatitlan group: (2) Minatitlan, Palestina* Olifantsvlei group: (3) Bobia, Botambi, Olifantsvlei Patois group: (6) Abras*, Babahoyo*, Pahayokee, Patois, Shark River, Zegla Simbu group: (25) Aino, Akabane, Buttonwillow, Douglas, Facey's Paddock*, Ingwavuma, Inini, Kaikalur, Manzanilla, Mermet, Nola, Oropouche, Peaton, Sabo, Sango, Sathuperi, Shamonda, Shuni, Simbu, Thimiri, Tinaroo, Utinga, Utive, Yaba-7* Tete group: (5) Bahig, Batama, Matruh, Tete, Tsuruse Turlock group: (6) Barmah Forest, Lednice*, M'Poko, Turlock, Umbre, Yaba-1 Ungrouped: (2) Enseada*, Kaeng Khoi	
Genus	Sandfly fever group	*Phlebovirus*
Type species	Sandfly fever (SF) Sicilian virus	–
Main characteristics	A. Properties of the Virus Particle Nucleic acid: L RNA, $2.6–2.8 \times 10^6$; M RNA, $1.8–2.2 \times 10^6$; S RNA, $0.7–0.8 \times 10^6$. Proteins: G1, $55–70 \times 10^3$; G2, $50–60 \times 10^3$; N, $20–30 \times 10^3$; L, $145–200 \times 10^3$. Both glycoproteins coded by M RNA; N coded by S RNA. L protein probably coded by L RNA.	

Taxonomic status	*English vernacular name*	*International name*
	B. Replication At least one nonstructural S RNA gene product synthesized in infected cells. C. Biological Aspects Host range: Various vertebrate species; primarily phlebotomines but occasional alternate arthropod species, mosquitoes and *Culicoides*.	
Other members (probable member indicated by asterisk)	Considered as a single serological group (at least 30 viruses). Serologically unrelated to members of other genera. The members are: Aguacate, Alenquer, Anhanga, Arumowot, Buenaventura, Bujaru, Cacao, Caimito, Candiru, Chagres, Chilibre, Frijoles, Gabek Forest, Gordil, Icoaraci, Itaituba, Itaporanga, Karimabad, Nique, Pacui, Punta Toro, Rift Valley fever, Rio Grande, Saint Floris*, Salehabad, SF-Naples, SF-Sicilian, Tehran, Toscana, Urucuri	
Genus	Nairobi sheep disease and related viruses	*Nairovirus*
Type species	Crimean-Congo hemorrhagic fever (CCHF), Crimean hemorrhagic fever (CHF) or Congo (CON) strains	–
Main characteristics	A. Properties of the Virus Particle Nucleic acid: L RNA, $4.1–4.9 \times 10^6$; M RNA, $1.5–1.9 \times 10^6$; S RNA, $0.6–0.7 \times 10^6$. Proteins: G1, $72–84 \times 10^3$; G2, $30–40 \times 10^3$; N, $48–54 \times 10^3$; L, $145–200 \times 10^3$. B. Replication At least two nonstructural glycopolypeptide precursors synthesized in infected cells. C. Biological Aspects Host range: Various vertebrate species; primarily ticks but occasional alternate arthropod species, mosquitoes and *Culicoides*.	
Other members (probable members indicated by asterisk)	Organized in six serogroups (at least 27 viruses). Serologically unrelated to members of other genera. The groups are: Crimean-Congo hemorrhagic fever group: (2) CCHF, Hazara Dera Ghazi Khan group: (6) Dera Ghazi Khan, Abu Hammad, Abu Mina, Kao Shuan, Pathum Thani, Pretoria Hughes group: (8) Hughes, Farallon, Fraser Point*, Punta Salinas, Raza*, Sapphire 11*, Soldado, Zirqa Nairobi sheep disease: (3) Nairobi sheep disease, Ganjam, Dugbe Qalyub: (2) Qalyub, Bandia Sakhalin group: (6) Avalon, Clo Mor, Paramushir*, Sakhalin, Taggert, Tillamook	
Genus	Uukuniemi and related viruses	*Uukuvirus*
Type species	Uukuniemi virus	–
Main characteristics	A. Properties of the Virus Particle Nucleic acid: L RNA, $2–2.5 \times 10^6$; M RNA, $1–1.3 \times 10^6$; S RNA, $0.4–0.6 \times 10^6$.	

Taxonomic status	English vernacular name	International name
	Proteins: G1, 70–75 × 10^3; G2, 65–70 × 10^3; N, 20–25 × 10^3; L, 180–200 × 10^3. Both glycoproteins coded by M RNA, synthesized as a precursor (p110); N coded by S RNA; L protein probably coded by L RNA. B. Replication At least one nonstructural polypeptide (NS, 30 × 10^3) synthesized in infected cells and coded by S RNA. C. Biological Aspects Host range: Various vertebrate species; ticks.	
Other members (probable members indicated by asterisk)	Organized in a single serogroup (at least 7 viruses). Serologically unrelated to members of other genera. The members are: Grand Arbaud, Manawa, Oceanside*, Ponteves, Uukuniemi, Zaliv-Terpeniya, EgAn-1825-61*	
Other possible members of family	At least 4 serogroups of viruses and some unassigned viruses. Serologically unrelated to members of other ***Bunyaviridae*** genera. The viruses are: Bakau group: (2) Bakau, Ketapang Kaisodi group: (3) Kaisodi, Lanjan, Silverwater Mapputta group: (4) Mapputta, Gan Gan, Maprik, Trubanaman Thogoto group: (2) SiAr126, Thogoto Unassigned viruses: (11) Belmont, Bhanja, Dhori, Khasan, Kowanyama, Lone Star, Razdan, Sunday Canyon, Tamdy, Tataguine, Witwatersrand	
Derivation of names	bunya: from *Bunya*mwera; place in Uganda, Africa, where type species was isolated nairo: from *Nairo*bi sheep disease; first reported disease that is caused by member virus phlebo: refers to *phlebo*tomine vectors; Greek *phlebos,* 'vein' uuku: from *Uuku*niemi; place in Finland where virus was isolated	

References

Berge, T.O.: International catalogue of arboviruses including certain other viruses of vertebrates. US Department of Health, Education and Welfare, DHEW Publ. No. (CDC) 75-8301 (1975).

Bishop, D.H.L.; Calisher, C.H.; Casals, J.; Chumakov, M.P.; Gaidomovich, S.Ya.; Hannoun, C.; Lvov, D.K.; Marshall, I.D.; Oker-Blom, N.; Pettersson, R.F.; Porterfield, J.S.; Russell, P.K.; Shope, R.E.; Westaway, E.G.: Bunyaviridae. Intervirology *14:* 125–143 (1980).

Bishop, D.H.L.; Shope, R.E.: Bunyaviridae; in Fraenkel-Conrat, Wagner, Comprehensive virology, vol. 14, pp. 1–156 (Plenum, New York 1979).

Karabatsos, N.: Supplement to international catalogue of arboviruses including certain other viruses of vertebrates. Am. J. trop. Med. Hyg. *27:* 372–440 (1978).

Montgomery, R.E.: On a tick-borne gastro-enteritis of sheep and goats occurring in British East Africa. J. comp. Path. Ther. *30:* 28–57 (1917).

Ushijima, H.; Klimas, R.; Kim, S.; Cash, P.; Bishop, D.H.L.: Characterization of the viral ribonucleic acids and structural polypeptides of Anopheles A, Bunyamwera, Group C, California, Capim, Guama, Patois, and Simbu bunyaviruses. Am. J. trop. Med. Hyg. *29:* 1441–1452 (1980).

Taxonomic status	*English vernacular name*	*International name*
Family	Arenavirus group	***ARENAVIRIDAE***

Main characteristics A. Properties of the Virus Particle

Nucleic acid: ssRNA; 5 RNAs repeatedly isolated – two virus-specific molecules L and S (distinctly different by oligonucleotide fingerprint analysis and lack of cross-hybridization with cDNAs) with apparent MWs of 1.1–1.6 and 2.1–3.2 × 10^6 and three of host cell origin with sedimentation coefficients of 28S and 18S (rRNAs) and 4–6S. The L and S RNAs of Pichinde virus contain similar termini of about 30 nucleotides at both the 5′ and 3′ termini. Some of the arenavirus DI particles appear to lack, or have diminished amounts of, L and S RNAs. They may contain 1–5 distinctly different RNA species. In certain viruses, closed circular as well as linear and hairpin forms have been observed along with considerable RNase-resistant structures.

Protein: One major nonglycosylated polypeptide tightly associated with the RNA with a MW between 63,000 and 72,000; thought to be part of the RNP complex and containing the cross-reacting antigens linking together this family of viruses. One or two major glycosylated polypeptide size classes; one with MW about 34–44 × 10^3 is consistently found, while the other with MW about 54–72 × 10^3 migrating close to the major nonglycosylated polypeptide may be found in some, but not all, viruses. The polypeptide profiles of DI particles of each arenavirus thus far examined are similar to those of the standard virus. Virion-associated enzymes: a transcriptase was found associated with the RNP complex of Pichinde virus. Poly(U) and poly(A) polymerases were also found with ribosomes; these latter two polymerases may be accidentally packaged during morphogenesis and may have no required role in virus replication.

Lipid: Present.

Carbohydrate: Glucosamine, fucose, galactose.

Physicochemical properties: MW not easily determined. $S_{20w} \simeq 325$–500. Density in sucrose$\simeq$1.17–1.18 g/cm^3; in CsCl = 1.19–1.20 g/cm^3; in amidotrizoate compounds = 1.14 g/cm^3. Relatively unstable in vitro. Rapidly inactivated below pH 5.5 and above pH 8.5. Inactivated rapidly at 56° and by solvents. Highly sensitive to UV and gamma radiation.

Morphology: Enveloped spherical to pleomorphic particles, 50–300 nm diameter (mean 110–130 nm). The dense lipid bilayer envelope has surface projections 10 nm long and club-shaped. Varying numbers of ribosome-like particles (20–25 nm diameter) appear free within the envelope.

Antigenic properties: Number of distinct antigenic molecules in virion is >3, based on evidence from biologic, serologic and pathologic data. Virion surface projections, but not inclusion body or supernatant antigens, are involved in virus neutralization. Antigens involved in neutralization are highly type-specific (although cross-neutralization tests demonstrate shared antigens between Tacaribe and Junin viruses). CF antigens are used to define the Tacaribe complex. The major CF antigen is in the RNP. Antigenic properties used for classification: fluorescent antibody techniques show that antisera against all Tacaribe complex viruses, as well as Lassa fever virus, react with lymphocytic choriomeningitis (LCM) virus. No hemagglutinin has been found yet. By monoclonal- or indirect fluorescent-antibody tests, Old World arenaviruses (LCM, Mozambique, and Lassa) are distinguishable from New World arenaviruses (the Tacaribe complex viruses). Monoclonal antibodies can also distinguish different Old World arenaviruses.

Effects of virus suspensions on cells: Most, if not all, arenaviruses probably have cell-killing potential. However, this is blunted in many types of infected cells by genesis of DI virus particles. Virus produced from cells which do not favor enrichment for DI particles, i.e.,

Taxonomic status	*English vernacular name*	*International name*

suspensions containing mostly standard virus, will probably induce transient CPE even in normally resistant cell types.

B. Replication

Molecular biology: Except where noted, information is based on studies with Pichinde virus. High-frequency genetic recombination found in two groups with temperature-sensitive mutants (VSV-Pichinde and VSV-Junin pseudotypes have also been found). Viral RNA transcribed by transcriptase into complementary RNA, which acts as mRNA. Virus-infected cells contain a 64,000 MW RNP and two other nonglycosylated peptides (MWs 42 and 200×10^3), the smaller giving rise to a fully glycosylated precursor of 79×10^3 MW (also found in LCM-infected cells) which in turn is cleaved to yield the two envelope glycopeptides. Genetic assignments, based on the use of intertypic recombinant viruses as well as calculated coding capacity of the RNAs, indicate that S RNA contains information for the nucleoprotein and precursor glycopeptide, while L RNA determines plaque morphology and the large nonglycosylated peptide. Recombinant studies with LCM virus have confirmed that L RNA controls plaque morphology. With LCM it was also possible to demonstrate that S RNA determines pathogenicity. The synthesis of LCM DI virus has been observed in vivo as well as in vitro. DNA synthesis inhibitors have no effect on arenavirus RNA synthesis, but a functional host-cell nucleus is required for virus multiplication. Replication in vitro of a number of arenaviruses is inhibited by amantadine, α-amanitin, benzimidazoles, glucosamine, and thiosemicarbazones; ribavirin appears to inhibit the replication of several arenaviruses in vitro and spares monkeys infected with Machupo and Lassa viruses.

Virus inclusion bodies: Intracytoplasmic inclusion bodies are prominent in cells infected with arenaviruses; they are made up of masses of ribosomes in a moderately electron-dense matrix. The relative proportion of ribosomes and matrix may vary widely in different inclusions, but as infection progresses a condensation of inclusion material results in rather uniformly marginated, large masses.

C. Biological Aspects

Host range: Natural – most viruses have a single restricted rodent host *(Mus, Calomys, Mastomys, Oryzomys, Sigmodon,* and the fruit-eating bat *Artibeus)* in which persistent infection with viremia and/or viruria occur, known or suspected to be caused by a slow and/or insufficient immune response of the host. Natural spread to other mammals and man is unusual; serial infection outside the rodent niche is limited. Experimental in vivo – Disease and outcome in laboratory animals (mouse, hamster, guinea pig, rhesus monkey, marmoset, rat) vary with the type of virus used. In general, viruses of the Tacaribe complex are pathogenic for suckling but not weaned mice; LCM and Lassa produce the opposite effect. Cross-protection is seen against Junin and Lassa with prior infection by Tacaribe and Mozambique viruses, respectively. LCM virus has been found to grow in murine lymphocytes. Experimental in vitro – Vero and BHK21 cells most commonly used for isolation and growth, but viruses grow moderately well in many other mammalian and *Hyalomma* cells.

Transmission: Vertical – transuterine, transovarian and postpartum (most likely by milk-, saliva-, or urine-borne routes) in natural host. Horizontal – Important as a mechanism for viruses to escape from their natural host. Venereal transmission suspected as an important mode for intra-species spread. Vectors – A few arthropod isolations which have never been shown to have any place in transmission cycles in nature. Biological – Unknown. Mechanical – Unknown.

Taxonomic status	English vernacular name	International name
Genus	LCM virus group	*Arenavirus*
Type species	Lymphocytic choriomeningitis virus	–
Other members	Lassa Tacaribe complex Amapari Junin Latino Machupo Parana Pichinde Tacaribe Tamiami	
Probable members	Flexal, BeAN (293022) Mozambique	
Derivation of name	arena: from Latin *arenosus,* 'sandy', from appearance of particles in electron microscope sections	

References

uchmeier, M.J.; Lewicki, H.A.; Tomori, O.; Johnson, K.M.: Monoclonal antibodies to lymphocytic choriomeningitis virus react with pathogenic arenaviruses. Nature, Lond. *288:* 486–487 (1980).

uchmeier, M.J.; Welsh, R.M.; Dutko, F.J.; Oldstone, M.B.A.: The virology and immunobiology of lymphocytic choriomeningitis virus infection. Adv. Immunol. *30:* 275–331 (1980).

uckley, S.M.; Casals, J.: Pathobiology of Lassa fever. Int. Rev. exp. Path. *18:* 97–136 (1978).

hrling, P.B.; Hesse, R.A.; Eddy, G.A.; Johnson, K.M.; Callis, R.T.; Stephen, E.: Lassa virus infection of rhesus monkeys: pathogenesis and treatment with ribavirin. J. infect. Dis. *141:* 580–589 (1980).

iley, M.P.; Lange, J.V.; Johnson, K.M.: Protection of rhesus monkeys from Lassa virus by immunization with closely related arenavirus. Lancet *ii:* 738 (1979).

irk, W.E.; Cash, P.; Peters, C.J.; Bishop, D.H.L.: Formation and characterization of an intertypic lymphocytic choriomeningitis recombinant virus. J. gen. Virol. *51:* 213–218 (1980).

ehmann-Grube, F.: Infection of mice with lymphocytic choriomeningitis virus; in Foster, Small, Fox, The mouse in biomedical research (Academic Press, New York, in press, 1981).

eung, W.-C.; Ramsingh, A.; Dimock, K.; Rawls, W.E.; Petrovich, J.; Leung, M.: Pichinde virus L and S RNAs contain unique sequences. J. Virol. *37:* 48–54 (1981).

ersich, S.E.; Leon, M.E.; Coto, C.E.: Cell nucleus participation in the multiplication of the arenavirus Tacaribe. FEMS Microbiol. Lett. *6:* 205–207 (1979).

onath, T.P.: International symposium on arenaviral infections of public health importance. Bull. Wld Hlth Org. *52:* 381–766 (1975).

ldstone, M.B.A.; Peters, C.J.: Arenavirus infections of the nervous system; in Klawans, Infections of the nervous system. Handbook of clinical neurology, vol. 34, pp. 193–207 (Elsevier/North Holland, Amsterdam 1978).

dersen, I.R.: Structural components and replication of arenaviruses. Adv. Virus Res. *24:* 277–330 (1979).

Pinheiro, F.P.; Woodall, J.P.; Travassos Da Rosa, A.P.A.; Travassos Da Rosa, J.F.: Studies on arenaviruses in Brazil. Medicina, B. Aires *37:* 175–181 (1977).

Rawls, W.E.; Leung, W.C.: Arenaviruses; in Fraenkel-Conrat, Wagner, Comprehensive virology, vol. 14, pp. 157–192 (Plenum, New York 1979).

Vezza, A.C.; Cash, P.; Jahrling, P.; Eddy, G.; Bishop, D.H.L.: Arenavirus recombination: the formation of recombinants between prototype Pichinde and Pichinde Munchique viruses and evidence that arenavirus S RNA codes for N polypeptide. Virology *106:* 250–260 (1980).

Taxonomic status	*English vernacular name*	*International name*
Group (monotypic)	Tomato spotted wilt virus group	–
Type species	Tomato spotted wilt virus (TSWV) (39)	–
Main characteristics	A. Properties of the Virus Particle Nucleic acid: ssRNA probably positive sense; probably four distinct molecules with MWs about 2.6, 1.9, 1.7 and 1.3×10^6. RNA is not infectious, but isolated RNP retains some infectivity. Protein: Four major polypeptides with MWs$\simeq$27, 52, 58 and 78×10^3, and up to three minor ones. The first one (MW 2×10^3) is located within the envelope. Lipid: In envelope, about 20% by weight of virus particle. Carbohydrate: One minor protein and all four major structural proteins are glycosylated. Physicochemical properties: $S_{20w} \simeq 560$; density$\simeq$1.21 g/cm^3 in sucrose; virus inactivated very rapidly in crude sap preparations, especially under oxidizing conditions. Morphology: Spherical, enveloped particles$\simeq$82 nm in diameter with an electron-dense layer external to the lipid bilayer. Internal RNP possibly in the form of tightly coiled strands. Antigenic properties: Relatively poor immunogens. Antisera react with all four major structural proteins. B. Replication Granular inclusions in cytoplasm of infected cells. Early in infection, individual immature particles are surrounded by two lipoprotein membranes. Later, mature particles having a single envelope occur in groups within smooth-membraned vesicles. Maturation resembles that of ***Bunyaviridae*** rather than ***Orthomyxoviridae***. C. Biological Aspects Host range: Very wide host range among plants. Transmission: Transmitted by thrips in a persistent manner; acquired only by larvae. Virus readily transmitted experimentally by sap inoculation.	

References

Milne, R.G.: An electron microscope study of tomato spotted wilt virus in sections of infected cells and in negative stain preparations. J. gen. Virol. *6:* 267–276 (1970).

Mohamed, N.A.: Isolation and characterization of subviral structures from tomato spotted wilt virus. J. gen. Virol. *53:* 197–206 (1981).

Tas, P.W.L.; Boerjan, M.L.; Peters, D.: The structural proteins of tomato spotted wilt virus. J. gen. Virol. *36:* 267–279 (1977).

van den Hurk, J.; Tas, P.W.L.; Peters, D.: The ribonucleic acid of tomato spotted wilt virus. J. gen. Virol. *36:* 81–91 (1977).

Taxonomic status	*English vernacular name*	*International name*
Family	RNA tumor viruses (and related agents)	***RETROVIRIDAE***

Main characteristics (most of the data are derived from the subfamily ***Oncovirinae***)

A. Properties of the Virus Particle

Nucleic acid: Inverted dimer of linear positive-sense ssRNA (about 2% by weight). Monomers held together at the 5′ ends by hydrogen bonds, probably base pairing. Non-defective viruses contain monomeric RNA of MW$\simeq 3 \times 10^6$. Polyadenylated at the 3′ end with a cap structure ($m^7G^5ppp^{5'}NmpNp$) at the 5′ end. Redundant sequences at the 5′ and 3′ ends. A cellular tRNA serving as primer for reverse transcription is bound by base pairing to a specific primer attachment site about 100 residues from the 5′ end of the genome. The virion RNA is not infectious.

Protein: About 60% by weight. Four internal nonglycosylated structural proteins [*gag* protein, two envelope *(env)* proteins, plus reverse transcriptase].

Lipid: About 35% by weight. Derived from the plasma membrane.

Carbohydrate: About 3.5% by weight. At least one of the two *env* proteins is glycosylated; in some viruses both are. Cellular carbohydrates and glycolipids are found in the viral envelope.

Physicochemical properties: Density between 1.16 and 1.18 g/cm^3 in sucrose gradients. Disrupted by lipid solvents and detergents. Surface glycoproteins partially removable by proteolytic enzymes. Half-life at 37° = 3–6 h. Relatively resistant to UV light.

Morphology: Spherical, enveloped virions 80–100 nm in diameter. Glycoprotein surface projections of approximately 8 nm diameter. Internal structure: probably icosahedral capsid containing a possibly helical RNP. Special features in thin sections: outer envelope, inner membrane (shell) and central nucleoid. The central nucleoid is located acentrically in type B ***Oncovirinae*** and concentrically in type C ***Oncovirinae.***

Antigenic properties: Virion proteins contain type-specific and group-specific determinants, the latter being shared among members of a genus. The type-specific determinants of the envelope glycoproteins are involved in antibody neutralization. Group-specific determinants of the major internal *gag* protein define subgenera, e.g., mammalian type C viruses.

Genetic structure: Although virions carry two copies of the genome, it is not known whether both are functional. Basic genetic information for the production of infectious progeny virus consists of 3 genes: *gag,* coding for internal nonglycosylated proteins of the virion; *pol,* coding for reverse transcriptase; and *env,* coding for envelope glycoproteins of the virion. Where the order of these genes has been determined, it reads 5′ *gag, pol, env* 3′. Some retroviruses also carry cell-derived genetic information for nonstructural proteins that are important in pathogenesis. These cellular sequences are either inserted in a complete retrovirus genome (some strains of Rous sarcoma virus) or they form substitutions for deleted viral replicative sequences (most other rapidly oncogenic retroviruses). Such deletions render the virus replication-defective and dependent on nontransforming helper virus for production of infectious progeny. In many cases the cell-derived sequences form a fused gene with viral structural information that is then translated into one protein (e.g., '*gag*-x' protein).

B. Replication

Entry into the host cell is mediated by interaction between an envelope glycoprotein of the virion and specific receptors at the cell surface, possibly resulting in fusion of the viral envelope to the plasma membrane. The further process of intracellular uncoating of the viral particle is not understood. Replication starts with reverse transcription of virion RNA

Taxonomic status	English vernacular name	International name

into DNA. The linear dsDNA transcripts of the viral genome contain large terminal repeats composed of sequences from the 3′ and 5′ ends of the viral RNA (3′5′– – – – 3′5′ structure). These repeats are also present in one form of circular viral DNA that occurs later in infection in the nucleus. A second circular form has only one 3′5′ repeat and may be derived from the first form by intramolecular recombination. The structure of viral DNA with its large terminal repeats resembles that of transposable genetic elements of procaryotes. Retroviral DNA becomes integrated into the chromosomal DNA of the host by an as yet unknown mechanism which, similar to integration of transposons, generates a short duplication of cell sequences at the integration site. Viral DNA can integrate at many sites in the cellular genome, but there is no evidence that once integrated it is transposed within the same cell. The map of the integrated provirus is coextensive with that of unintegrated linear viral DNA. Integration appears to be a prerequisite of virus replication. Integrated provirus is transcribed by cellular RNA polymerase II into virion RNA and mRNA. There are several classes of mRNA reflecting the genetic map of retroviruses. An mRNA comprising the whole genome serves for the translation of the *gag* and *pol* genes positioned at the 5′ portion of this RNA into structural proteins and reverse transcriptase, respectively. A smaller mRNA consisting of the 3′ sequences of the genome, including the *env* gene and the C region, is translated into the precursor of the envelope proteins. All mRNAs share a common sequence at their 5′ end, and in the less-than-genome size mRNAs this sequence appears to be acquired by RNA splicing. Most primary translational products in retrovirus infection are polyproteins which require proteolytic cleavage before becoming functional. Virus matures at the plasma membrane and is released from the cell by a budding process.

C. Biological Aspects
Host range: Endogenous oncoviruses that are part of the germ line and are inherited as Mendelian genes occur widely among vertebrates. Association with diseases: leukemias, lymphomas, sarcomas and other tumors of mesodermal origin, mammary carcinomas, carcinomas of liver and kidney, autoimmune disease, lower motor neuron disease, several acute diseases with tissue damage. Some retroviruses may be nonpathogenic.
Transmission: Transmission is vertical and horizontal.

Taxonomic status	English vernacular name	International name
Subfamilies	RNA tumor virus group	***Oncovirinae***
	Foamy virus group	***Spumavirinae***
	Maedi/visna group	***Lentivirinae***
Subfamily	RNA tumor virus group	***Oncovirinae***
Genera	Type C oncovirus group	–
	Type B oncovirus group	–
Proposed genus	Type D retrovirus group	–
Genus	Type C oncovirus group	–
Subgenera	Mammalian type C oncoviruses	–
	Avian type C oncoviruses	–
	Reptilian type C oncoviruses	–

Taxonomic status	*English vernacular name*	*International name*
Subgenus	Mammalian type C oncoviruses	–
Species	Baboon type C oncovirus Bovine leukosis virus Feline sarcoma and leukemia viruses Gibbon ape leukemia virus Guinea pig type C oncovirus Murine sarcoma and leukemia viruses Porcine type C oncovirus Rat type C oncovirus (rat leukemia) Woolly monkey sarcoma virus	
Subgenus	Avian type C oncoviruses	–
Species	Avian reticuloendotheliosis virus Avian sarcoma and leukemia viruses Pheasant type C oncoviruses	
Subgenus	Reptilian type C oncoviruses	–
Species	Viper type C oncovirus	
Genus	Type B oncovirus group	–
Species	Mouse mammary tumor viruses	
Proposed genus	Type D retrovirus group	–
Species	Mason-Pfizer monkey virus	
Subfamily	Foamy virus group	***Spumavirinae***

Main characteristics

A. Properties of the Virus Particle

Spumaviruses have many of the morphological, physical and chemical characteristics of other retroviruses (60–70S RNA, reverse transcriptase). Distinctive features include: electron-lucent nucleoids, long spikes projecting from the surface, maturation by budding into intracytoplasmic vacuoles. Serotypes are identified by neutralization and CF tests; with few exceptions the neutralization test is type-specific.

B. Replication

Spumaviruses replicate only in dividing cells. Early events are similar to other enveloped viruses. After uncoating, genetic information is transferred to a DNA intermediate. Infectious DNA can be isolated from infected cells, but little is known of its synthesis, size or structure. Viral DNA serves as template for synthesis of RNA and virus polypeptides which are first detected by immunofluorescence in the nucleus, in contrast to other retroviruses. Later, viral antigens are found in cytoplasm and plasmia membrane. A single cycle of replication is completed in 5–7 days, with a latency period of 24 h.

Taxonomic status	English vernacular name	International name

C. Biological Aspects

Spumaviruses infect a variety of mammalian species (primates, possibly man, hamsters, cats, cows, rabbits). The salient features of infection by these agents are (i) persistent infection in the animal despite the host immune response, (ii) absence of pathological damage, (iii) 'unmasking' of infection in vitro with spontaneous formation of foamy vacuolated syncytia.

Spumaviruses do not induce transformation of cells in vitro, or cause tumors in animals.

Species	Bovine syncytial Feline syncytial Human foamy Simian foamy	

Subfamily	Maedi/visna group	***Lentivirinae***

Main characteristics

A. Properties of the Virus Particle

Visna virus, the best-studied member of the lentiviruses, is similar to type C oncoviruses in morphology, physical properties and chemical composition. Surface projections of the virus are composed of a glycoprotein (gp135) that induces type-specific neutralizing antibody. There are three major internal structural polypeptides – p30, p16, and p14; p30 bears group-specific antigenic determinants that are shared by other members of the subfamily (Maedi, progressive pneumonia, and Zwogerziekte viruses), but not with oncoviruses or spumaviruses. The nucleoid contains reverse transcriptase, a dimer of subunits of MW$\simeq 68 \times 10^3$. The genome is a linear positive-sense ssRNA which is not infectious. The monomeric RNA has a MW of 3.5×10^6 and is polyadenylated at the 3′ end. In the virion, three monomers may be hydrogen-bonded to form concentric circles. A cellular tRNA serving as primer for reverse transcription is base-paired to genomic RNA.

B. Replication

Lentiviruses probably enter the cell by fusion of the viral envelope to the plasma membrane, and RNA is released into the cytoplasm into a vacuolar structure. Replication begins in the cytoplasm with reverse transcription, but DNA synthesis is completed in the nucleus. The predominant DNA species in the nucleus is a gapped structure composed of a full-length minus strand colinear with virion RNA and two long plus strands with an intervening central gap of about 500 nucleotides. Completed molecules are linear duplexes that bear terminal repeats of about 400 nucleotides. Two circular forms differing by the length of 1 terminal repeat constitute a minor proportion. Under permissive conditions DNA synthesis continues throughout the infectious cycle to reach levels of 300–500 genome equivalents per cell. Synthesis occurs in two phases. The second phase is due to superinfection and generates DNA that is not required for a full yield of virus. In the first phase during the latency period most of the incoming RNA is transcribed into DNA, and the extent of DNA synthesis in this phase regulates the subsequent rate and extent of production of RNA and progeny virus. A variable proportion of viral DNA in the nucleus is associated with a high-molecular-weight cellular DNA, but it has not been established that the linkage is covalent or that integration is required for replication. There are three size classes of mRNA but their relationship to specific genes is not known. Similarly, the primary translation products are polyproteins, but details of the subsequent cleavage events have not been determined.

Taxonomic status	*English vernacular name*	*International name*
	C. Biological Aspects Host range: Lentiviruses are horizontally transmitted in infections limited to sheep. No evidence exists for endogenous lentiviruses or endogenous viral genes. Association with diseases: Lentiviruses cause destructive inflammatory pathology in the lungs and central nervous system of sheep which evolves slowly over months to many years, characteristic of slow infections.	
Species	Maedi Progressive pneumonia Visna Zwogerziekte	
Derivation of names	retro: Latin 'backwards' (refers also to reverse transcriptase) onco: from Greek *onkos,* 'tumor' spuma: Latin 'foam' lenti: Latin 'slow'	

References

Bishop, J.M.: Retroviruses. A. Rev. Biochem. *47:* 35–88 (1978).

Bishop, J.M.: Enemies within: the genesis of retrovirus oncogenes. Cell *23:* 5–6 (1981).

Fine, D.; Schochetman, G.: Type D primate retroviruses: a review. Cancer Res. *38:* 3123–3139 (1978).

Haase, A.T.: The slow infection caused by Visna virus. Curr. Top. Microbiol. Immunol. *71:* 101–156 (1975).

Hanafusa, H.: Cell transformation by RNA tumor viruses; in Fraenkel-Conrat, Wagner, Comprehensive virology, vol. 10, pp. 401–483 (Plenum, New York 1977).

Hooks, J.J.; Gibbs, C.J., Jr.: The foamy viruses. Bact. Rev. *39:* 169–185 (1975).

Stephenson, J. (ed.): Molecular biology of RNA tumor viruses (Academic Press, New York 1980).

Tooze, J. (ed.): Molecular biology of tumor viruses; 2nd ed., part 1 (Cold Spring Harbor Press, Cold Spring Harbor 1982).

Vogt, P.K.: Genetics of RNA tumor viruses; in Fraenkel-Conrat, Wagner, Comprehensive virology, vol. 9, pp. 341–455 (Plenum, New York 1977).

Vogt, P.K.; Ju, S.S.F.: The genetic structure of RNA tumor viruses. A. Rev. Genet. *11:* 203–238 (1977).

Taxonomic status	*English vernacular name*	*International name*
Family	Picornavirus group	***PICORNAVIRIDAE***

Main characteristics A. Properties of the Virus Particle

Nucleic acid: One molecule of infectious positive-sense ssRNA, MW$\simeq$2.5 × 10^6. A polyadenylic sequence is transcribed onto the 3′ terminus; there is a covalently linked protein, VPg (MW$\simeq$2,400), at the 5′ terminus.

Protein: Four major polypeptides (60 molecules of each per virion): three of MW 24–41 × 10^3, one of MW 5.5–13.5 × 10^3.

Lipid: None.

Carbohydrate: None.

Physicochemical properties: MW 8–9 × 10^6; S_{20w} = 140–165; buoyant density in CsCl = 1.33–1.45 g/cm^3 depending mainly on genus. Some species have infectious particles of two different densities. Virus particles are generally not disrupted by neutral detergents. Some species are unstable below pH 6, in some cases depending on the ionic environment. Inactivated by photodynamic action in the presence of certain dyes.

Morphology: Virus particles are roughly spherical naked nucleocapsids, 22–30 nm in diameter (icosahedral with T = 1). There is no envelope or core; there are no surface projections, the surface usually being almost featureless. The capsid consists of 60 subunits, each comprising 1 molecule of each of the four major capsid polypeptides, and formed by cleavage of 1 molecule of the structural protein precursor (MW 80–100 × 10^3). Some subunits may be incompletely cleaved.

Antigenic properties: Native virions are antigenically specific (designated 'N' or 'D'), but after denaturation are converted to group specificity (designated 'H'). Neutralization by antibody follows first-order inactivation kinetics. Species are classified by neutralization of infectivity, complement-fixation or immunodiffusion; some species can be identified by hemagglutination.

B. Replication

Replication of viral RNA occurs in complexes associated with cytoplasmic membranes apparently via two distinct partially dsRNA RIs; one complex uses positive-strand RNA and the other negative-strand RNA as template. Functional proteins are mainly produced by translation of a large (MW 210,000) precursor polyprotein from positive-strand RNA, followed by post-translational cleavage into functional polypeptides. Functions known to be virus-coded include coat protein (5′ half of message) and VPg, protease and polymerases or polymerase factors (3′ half). Many compounds specifically inhibiting replication have been described. Resistant and dependent mutants can be obtained. Genetic recombination, complementation and phenotypic mixing occur, as do DI particles.

C. Biological Aspects

Host range: Natural – except for coxsackie B5 virus, EMC virus and aphthoviruses (which infect most cloven-foot animals), most species are host-species specific. Artificial – most species can be grown in cell cultures. Resistant host cells can often be infected (single round) by transfection with naked infective RNA. Rhinoviruses and many enteroviruses grow poorly or not at all in laboratory animals.

Transmission: Horizontal, mainly mechanically.

Genera	Enterovirus	*Enterovirus*
	EMC virus group	*Cardiovirus*

Taxonomic status	*English vernacular name*	*International name*
	Common cold virus Foot-and-mouth disease virus	*Rhinovirus* *Aphthovirus*
Genus	Enterovirus	*Enterovirus*
Type species	Human poliovirus 1	–
Main characteristics	Stable at acid pH; buoyant density in CsCl = 1.33–1.34 g/cm^3; primarily viruses of gastro-intestinal tract, but also multiply in other tissues such as nerve, muscle, etc.	
Other members	Human polioviruses 2–3 Human coxsackieviruses A1–22, 24 (A23 = echovirus 9) Human coxsackieviruses B1–6 (swine vesicular disease virus is very similar to coxsackie-virus B5) Human echoviruses 1–9, 11–27, 29–34 Human enteroviruses 68–71 Human enterovirus 72 (hepatitis A virus) Murine poliovirus (Theiler's encephalomyelitis virus, TO, FA, GD VII) Simian enteroviruses 1–18 Porcine enteroviruses 1–8 Bovine enteroviruses 1–7	
Genus	EMC virus group	*Cardiovirus*
Type species	Encephalomyocarditis (EMC) virus	–
Main characteristics	Unstable at pH 5–6 in presence of 0.1 *M* halide; buoyant density in CsCl = 1.33–1.34 g/cm^3; unique serotype; clinical manifestations. Poly(C) tract of variable length (80–250 bases) about 150 bases from 5′ terminus of RNA.	
Other members	Mengovirus Murine encephalomyelitis (ME)	
Genus	Common cold virus	*Rhinovirus*
Type species	Human rhinovirus 1A	–
Main characteristics	Unstable below pH 5–6; buoyant density in CsCl = 1.38–1.42 g/cm^3; clinical manifestations	
Other members	Human rhinoviruses 2–113 Bovine rhinoviruses 1 and 2	
Genus	Foot-and-mouth disease virus	*Aphthovirus*
Type species	Aphthovirus O	–

Taxonomic status	*English vernacular name*	*International name*
Main characteristics	Unstable below pH 5–6; buoyant density in CsCl = 1.43–1.45 g/cm^3; clinical manifestations. Poly(C) tract of variable length (100–170 bases), about 400 bases from 5′ terminus of RNA.	
Other members	A C SAT1 SAT2 SAT3 Asia 1	
Other members of family ***Picornaviridae*** not yet assigned to genera	Cricket paralysis virus *Drosophila* C virus Equine rhinoviruses types 1 and 2 *Gonometa* virus	
Unclassified small RNA viruses of invertebrates	About 30 small RNA viruses of unknown affinities have been described. These include: bee acute paralysis, bee slow paralysis, bee virus X, *Drosophila* P and A, and sacbrood viruses.	
Derivation of names	picorna: from the prefix 'pico' (='micro-micro') and RNA (=the sigla for ribonucleic acid). entero: from Greek *enteron,* 'intestine' rhino: from Greek *rhis, rhinos,* 'nose' cardio: from Greek *kardia,* 'heart' aphtho: Greek *aphtha,* 'vesicles in the mouth'; English *aphtha,* 'thrush'; French *fièvre aphteuse*	

References

Agol, V.I.: Structure, translation and replication of picornaviral genomes. Prog. med. Virol., vol. 26, pp. 119–157 (Karger, Basel 1980).

Ahlquist, P.; Kaesberg, P.: Determination of the length distribution of poly(A) at the 3′-terminus of the virion RNAs of EMC virus, poliovirus, rhinovirus 14 and RAV-61. Nucl. Acids Res. *7:* 1195–1204 (1979).

Black, D.N.; Stephanson, P.; Rowlands, D.J.; Brown, F.: Sequence and location of the poly(C) tract in aphtho- and cardiovirus RNA. Nucl. Acids Res. *6:* 2381–2390 (1979).

Cooper, P.D.; Agol, V.I.; Bachrach, H.L.; Brown, F.; Ghendon, Y.; Gibbs, A.J.; Gillespie, J.H.; Lonberg-Holm, K.; Mandel, B.; Melnick, J.L.; Mohanty, S.B.; Povey, R.C.; Rueckert, R.R.; Schaffer, F.L.; Tyrrell, D.A.J.: Picornaviridae: second report. Intervirology *10:* 165–180 (1978).

Kitamura, N.; Semler, B.L.; Rothberg, P.G.; Larsen, G.R.; Adler, C.J.; Dorner, A.J.; Emini, E.A.; Hanecak, R.; Lee, J.J.; van der Werf, S.; Anderson, C.W.; Wimmer, E.: Primary structure, gene organization and polypeptide expression of poliovirus RNA. Nature, Lond. *291:* 547–553 (1981).

Knowles, N.J.; Buckley, L.S.: Differentiation of porcine enterovirus serotypes by complement fixation. Res. vet. Sci. *29:* 113–115 (1980).

Lipton, H.L.; Friedmann, A.: Purification of Theiler's murine encephalomyelitis virus. J. Virol. *33:* 1165–1172 (1980).

Melnick, J.L.; Tagaya, I.; von Magnus, H.: Enteroviruses 69, 70 and 71. Intervirology *4:* 369–370 (1974).

Rueckert, R.R.; Palmenberg, A.C.; Pallansch, M.: Evidence for a self-cleaving precursor of virus-coded protease, RNA replicase and VPg; in Koch, Richter, Biosynthesis, modification and processing of cellular and viral polyproteins, pp. 263–275 (Academic Press, New York 1980).

Sangar, D.V.: The replication of picornaviruses. J. gen. Virol. *45:* 1–13 (1979).

Scotti, P.D.; Longworth, J.F.; Plus, N.; Croizier, G.: Biology and ecology of strains of an insect small RNA virus complex. Adv. Virus Res. *26:* 117–143 (1981).

Taxonomic status	*English vernacular name*	*International name*
Family	Calicivirus family	***CALICIVIRIDAE***

Main characteristics A. Properties of the Virus Particle

Nucleic acid: One molecule of infectious positive-sense ssRNA, MW $2.6–2.8 \times 10^6$. Polyadenylated, probably at 3′ terminus; no methylated cap at 5′ terminus.

Protein: One major capsid polypeptide, MW $60–71 \times 10^3$; blocked N-terminal. A minor polypeptide, MW$\simeq 15 \times 10^3$ and $< 2\%$ of total protein, has also been observed. A protein with apparent MW$\simeq 10–15 \times 10^3$, essential for infectivity, is covalently linked to virion RNA, presumably at the 5′ end.

Lipid: None.

Carbohydrate: None reported.

Physicochemical properties: Virion MW$\simeq 15 \times 10^6$. Virions sediment at 170–183S. Buoyant density in neutral CsCl = 1.36–1.39 g/cm^3 depending upon virus strain. Not disrupted by ether, chloroform or mild detergents. Inactivated at pH values between 3 and 5. Thermal inactivation is accelerated in high concentrations of Mg^{++}. Some members inactivated by trypsin.

Morphology: Roughly spherical, 35–39 nm diameter, with 32 cup-shaped surface depressions arranged in icosahedral symmetry. Capsid probably consists of 180 polypeptides.

Antigenic properties: Neutralization indicates distinct serotypes of VESV and SMSV, but considerable cross-reactivity among FCV strains (see below for full virus names). Precipitin reactions indicate antigenic relationships among VESV, SMSV and FCV.

Effect on cells: Lysis.

B. Replication

Partially dsRNA (presumptive RI), dsRNA, genome-sized ssRNA and two smaller ssRNAs (MWs = 1.1 and 0.7×10^6) are found in infected cells. The 1.1×10^6 RNA is probably a subgenomic mRNA coding for the major capsid polypeptide (probably via cleavage of a precursor). Capsid polypeptide is the major protein synthetic product; an uncertain number of additional polypeptides is also synthesized; precursor-product relationships among them not fully established. Virions mature in cytoplasm.

C. Biological Aspects

Host range: Natural – vesicular exanthema of swine virus (VESV): swine (pinnipeds?); San Miguel sea lion virus (SMSV): pinnipeds, fish, swine; feline calicivirus (FCV): felines, dogs. Artificial – VESV in vivo: horse (some strains); SMSV in vivo: monkey. Cell culture – VESV/SMSV: porcine, primate (feline?); FCV: feline (primate?).

Transmission: Biological vectors not known. Mechanical via contaminated food (VESV), contact, airborne (FCV). Marine/terrestrial transmission likely with SMSV/VESV.

Genus	Calicivirus group	*Calicivirus*
Type species	Vesicular exanthema of swine virus (VESV) (serotype A)	–
Other members	VESV, approximately 12 additional serotypes Feline calicivirus (FCV, formerly called feline picornavirus), numerous antigenically related strains San Miguel sea lion (SMSV), 8 or more serotypes	

Taxonomic status	*English vernacular name*	*International name*
Possible members of ***Caliciviridae,*** genus uncertain	Amyelois chronic stunt virus Norwalk virus Viruses with calicivirus morphology that cause gastroenteritis in humans, calves, swine	
Derivation of name	calici: from Latin *calix,* 'cup' or 'goblet', from cup-shaped depressions observed by electron microscopy	

References

Greenberg, H.B.; Wyatt, R.G.; Kalica, A.R.; Yolken, R.H.; Black, R.; Kapikian, A.Z.; Chanock, R.M.: New insights in viral gastroenteritis. Perspect. Virol. *11:* 163–188 (1981).

Hillman, B.; Morris, T.J.; Kellen, W.R.; Hoffman, D.; Schlegel, D.E.: An invertebrate calici-like virus: evidence for partial virion disintegration in host excreta. (In preparation.)

Schaffer, F.L.: Caliciviruses; in Fraenkel-Conrat, Wagner, Comprehensive virology, vol. 14, pp. 249–284 (Plenum, New York 1979).

Schaffer, F.L.; Bachrach, H.L.; Brown, F.; Gillespie, J.H.; Burroughs, J.N.; Madin, S.H.; Madeley, C.R.; Povey, R.C.; Scott, F.; Smith, A.W.; Studdert, M.J.: Caliciviridae. Intervirology *14:* 1–6 (1980).

Studdert, M.J.: Caliciviruses: a brief review. Archs Virol. *58:* 157–191 (1978).

Taxonomic status	*English vernacular name*	*International name*
Family	*Nudaurelia* β virus group	–
Main characteristics	A. Properties of the Virus Particle Nucleic acid: One molecule of ssRNA. MW = 1.8×10^6; 10–11% of particle by weight. Protein: One major polypeptide of MW $60–70 \times 10^3$. Lipid: Not determined; probably none. Carbohydrate: Not determined. Physicochemical properties: MW = 16.3×10^6; $S_{20w} = 194–210$; buoyant density in CsCl = 1.275–1.298 g/cm^3. Stable at pH 3.0. Morphology: Virus particles are roughly spherical naked nucleocapsids 35 nm diameter, probably icosahedral with 240 subunits and T = 4. Antigenic properties: Most of the members of the group are serologically interrelated but distinguishable. B. Replication The viruses replicate primarily in the cytoplasm of gut cells of several Lepidoptera. C. Biological Aspects Host range: Natural – All species were isolated from Lepidoptera, principally from Saturniid, Limacodid and Noctuid moths. Artificial – No infections have yet been achieved in cultured invertebrate cells.	
Type species	*Nudaurelia* β virus (isolated from *Nudaurelia cytherea capensis*)	–
Main characteristics	As for the family description.	
Other members	Other viruses have been isolated from: *Antheraea eucalypti* *Darna trima* *Thosea asigna* *Philosamia ricini* *Trichoplusia ni*	

References

Finch, J.T.; Crowther, R.A.; Hendry, D.A.; Struthers, J.K.: The structure of *Nudaurelia capensis* β virus: the first example of a capsid with icosahedral surface symmetry T = 4. J. gen. Virol. *24:* 191–200 (1974).

Reinganum, C.; Robertson, J.S.; Tinsley, T.W.: A new group of RNA viruses from insects. J. gen. Virol. *40:* 195–202 (1978).

Struthers, J.K.; Hendry, D.A.: Studies of the protein and nucleic acid components of *Nudaurelia capensis* β virus. J. gen. Virol. *22:* 355–362 (1974).

Taxonomic status	*English vernacular name*	*International name*
Family	ssRNA phages	***LEVIVIRIDAE***
Genus	ssRNA phages	*Levivirus*
Type species	Phage MS2	–

Main characteristics A. Properties of the Virus Particle

Nucleic acid: One molecule of linear positive-sense ssRNA; MW$\simeq 1.2 \times 10^6$; 30% by weight of particle; 51–52% G+C content.

Protein: 180 copies of capsid protein (MW $12–14 \times 10^3$) and one copy of A protein (MW $35–44 \times 10^3$), which is required for maturation and infectivity. Capsid protein lacks histidine.

Lipid: None.

Carbohydrate: None.

Physicochemical properties: MW$\simeq 3.9 \times 10^6$; S_{20w} = 78–82; buoyant density in CsCl = 1.46 g/cm^3. Infectivity is ether- and chloroform-resistant.

Morphology: Virions are icosahedral, about 23 nm in diameter, and probably have 32 capsomers (T=3). No envelope.

B. Replication

Adsorption to sides of pili of male bacteria. Infecting phage RNA codes for a negative strand which acts as template; formation of partially dsRIs. Genome has 4 genes (for A protein, coat protein, lysis protein and RNA polymerase) and acts as messenger for protein synthesis. Phages form crystalline arrays in infected bacteria. On lysis of host cell, thousands of virus particles may be released.

C. Biological Aspects

Host range: Enterobacteria, *Caulobacter, Pseudomonas (Bdellovibrio?)*.

Other members

a. B6, B7, fcan, FH5, fr, f2, M12, Qβ, R17, R23, R34, ZG/1, ZIK/1, ZJ/1, ZL/3, ZS/3, α15, β, μ2, others (enterobacteria, 3 or 4 subgroups)
b. φCb5, φCb8r, φCb12r, φCb23r, φCP2, φCP18, φCr14, φCr28 *(Caulobacter)*
c. PPR1, PP7, 7s *(Pseudomonas)*

Possible members Viruses of *Bdellovibrio*

Derivation of name levi: from Latin *levis*, 'light'

References

Ackermann, H.-W.: Cubic, filamentous, and pleomorphic bacteriophages; in Laskin, Lechevalier, CRC handbook of microbiology; 2nd ed., vol. II, pp. 673–682 (CRC Press, West Palm Beach 1978).

Fiers, W.: Structure and function of RNA bacteriophages; in Fraenkel-Conrat, Wagner, Comprehensive virology, vol. 13, pp. 69–204 (Plenum, New York 1979).

Zinder, N.Z. (ed.): RNA phages (Cold Spring Harbor Laboratory, Cold Spring Harbor 1975).

Taxonomic status	*English vernacular name*	*International name*
Group	Maize chlorotic dwarf virus group	–
Type member	Maize chlorotic dwarf virus (MCDV) (194)	–
Main characteristics	A. Properties of the Virus Particle Nucleic acid: One molecule of positive-sense ssRNA, MW$\simeq$3.2 × 10^6. Percent base composition: G24, A30, C17, U29. Protein: Not characterized. Lipid: None reported. Carbohydrate: None reported. Physicochemical properties: MW$\simeq$8.8 × 10^6; $S_{20w}$$\simeq$183; density in CsCl = 1.51 g/cm^3. Morphology: Polyhedral particles $\simeq$30 nm diameter. Antigenic properties: Efficient immunogen. B. Replication Not known. C. Biological Aspects Host range: Narrow, limited to members of Gramineae. Transmission: Only by leafhoppers in a semi-persistent manner.	
Possible member	Rice tungro (67)	

References

Gingery, R.E.: Properties of maize chlorotic dwarf virus and its ribonucleic acid. Virology *73:* 311–318 (1976).
Gingery, R.E.; Gordon, D.T.; Nault, L.R.; Pradfute, O.E.: Maize chlorotic dwarf virus; in Kurstak, Handbook of plant virus infections and comparative diagnosis, pp. 19–32 (Elsevier/North Holland, Amsterdam 1981).

Taxonomic status	*English vernacular name*	*International name*
Group	Turnip yellow mosaic virus group	***TYMOVIRUS*** **(214)**
Type member	Turnip yellow mosaic virus (TYMV) (2, 230)	–

Main characteristics

A. Properties of the Virus Particle

Nucleic acid: One molecule of linear positive-sense ssRNA; MW$\simeq 2 \times 10^6$, accounting for $\simeq$35% of the weight of the B component. The 5′ terminus of TYMV RNA has the sequence $m^7G^{5\prime}ppp^{5\prime}Gp\ldots$; the 3′ terminus has a tRNA-like structure which accepts valine. Small amounts of subgenomic coat protein mRNA (MW$\simeq 0.2$–0.3×10^6) are found in several classes of virus particles. Both RNAs have a high content of cytidylic acid (32–41%). Particles of some members may also contain small amounts of transfer RNAs of plant origin.

Protein: One coat polypeptide, MW$\simeq$20,000. 180 molecules per particle.

Lipid: None.

Carbohydrate: None.

Physicochemical properties: Two major classes of stable particles (B and T) with MWs of 5.6 and 3.6×10^6; buoyant densities in CsCl$\simeq$1.42 and 1.29 g/cm^3, and $S_{20w} = 115$ and 54, respectively. Only the B component containing the genome RNA is infectious. Partial specific volume = 0.661. The isoelectric point of TYMV is pH 3.75; those of other members cover a wide pH range. Virus is stable at neutral pH. Several minor nucleoproteins can be isolated with densities intermediate between those of the two major particle types. For TYMV, these contain subgenomic coat protein mRNA and less than full-length pieces of the genome RNA.

Morphology: Particles are T=3 icosahedral structures, $\simeq$29 nm in diameter. They are stabilized by protein-protein interactions of the 180 subunits, which are clustered into 12 pentamers and 20 hexamers.

Antigenic properties: Serological relationships between members of the group range from very close, to distant, to not detectable.

B. Replication

Tymoviruses induce at the periphery of the chloroplasts small flask-shaped double-membrane bounded vesicles which contain membrane-bound viral polymerase. They are probably the main site of production of viral positive-sense RNA. Pentamers and hexamers of the protein are probably produced in the cytoplasm, and virions assembled at the necks of the vesicles. Empty protein shells accumulate in nuclei. Most members cause clumping of chloroplasts in infected cells.

C. Biological Aspects

Host range: Possibly restricted to dicotyledonous hosts. Individual viruses often have narrow host range.

Transmission: Mechanical and by beetles

Other members

Andean potato latent
Belladonna mottle (52)
Cacao yellow mosaic (11)
Clitoria yellow vein (171)
Desmodium yellow mottle (168)
Dulcamara mottle
Eggplant mosaic (124)

Taxonomic status	*English vernacular name*	*International name*
	Erysimum latent (222)	
	Kennedya yellow mosaic (193)	
	Okra mosaic (128)	
	Ononis yellow mosaic	
	Peanut yellow mottle	
	Physalis mosaic	
	Plantago mottle	
	Scrophularia mottle (113)	
	Wild cucumber mosaic (105)	
Possible member	Poinsettia mosaic virus	
Derivation of name	tymo: sigla from *t*urnip *y*ellow *mo*saic virus	

References

Fulton, R.W.; Fulton, J.L.: Characterization of a tymo-like virus common in Poinsettia. Phytopathology *70:* 321–324 (1980).

Koenig, R.: A loop-structure in the serological classification system of tymoviruses. Virology *72:* 1–5 (1976).

Koenig, R.; Lesemann, D.-E.: Tymoviruses; in Kurstak, Plant virus infections and comparative diagnosis, pp. 33–60 (Elsevier/North Holland, Amsterdam 1981).

Lana, A.F.: Properties of a virus occurring in *Arachis hypogea* in Nigeria. Phytopath. Z. *97:* 169–178 (1980).

Lesemann, D.-E.: Virus group-specific cytological alterations induced by members of the tymovirus group. Phytopath. Z. *90:* 315–336 (1977).

Matthews, R.E.F.: Tymovirus (turnip yellow mosaic virus) group; in Maramorosch, The atlas of insect and plant viruses, pp. 347–361 (Academic Press, New York 1977).

Szybiak, U.; Bouley, J.P.; Frisch, C.: Evidence for the existence of a coat protein messenger RNA associated with the top component of each of three tymoviruses. Nucl. Acids Res. *5:* 1821–1831 (1978).

Taxonomic status	*English vernacular name*	*International name*
Group	Barley yellow dwarf virus group	***LUTEOVIRUS***
Type member	Barley yellow dwarf virus (BYDV) – MAV isolate (32)	–
Main characteristics	A. Properties of the Virus Particle Nucleic acid: One molecule of positive-sense ssRNA. $MW \simeq 2.0 \times 10^6$. Protein: One coat polypeptide of $MW \simeq 24 \times 10^3$. Lipid: None reported. Carbohydrate: None reported. Physicochemical properties: $S_{20w} \simeq 115–127$. Usually resistant to chloroform. Morphology: Isometric particles, about 25–30 nm diameter. Antigenic properties: Strongly immunogenic. Most members are serologically related. B. Replication Confined to phloem tissue of infected plants. Details of ultrastructural changes vary among members. C. Biological Aspects Host range: Varies with member – some infect wide range of monocotyledonous plants, others infect many dicotyledonous plants, and some are restricted to smaller plant groups. Transmission: Persistent transmission by aphid vectors; virus apparently does not replicate in vector. Pronounced vector specificity among some virus isolates. Not transmitted by mechanical inoculation to plants, but aphids rendered inoculative by injection. Several members are associated with systems of dependent virus transmission by aphids from mixed infections in the host plant.	
Other members	Characterized isolates of BYDV fall into two groups on the basis of serological properties and cytological effects: I. MAV, PAV, and SGV II. RPV, RMV and rice giallume Beet mild yellowing Beet western yellows (89) Carrot red leaf Indonesian soybean dwarf Legume yellows *Malva* yellows Pea leaf roll (bean leaf roll) Potato leaf roll (36) *Solanum* yellows Soybean dwarf (179) Subterranean clover red leaf Tobacco necrotic dwarf (234) Turnip yellows	
Probable members	Milk-vetch dwarf Subterranean clover stunt Tomato yellow top	

Taxonomic status	*English vernacular name*	*International name*
Possible members	Banana bunchy top	
	Beet yellow net	
	Celery yellow spot	
	Cotton anthocyanosis	
	Filaree red leaf	
	Grapevine ajinashika	
	Groundnut rosette assistor	
	Millet red leaf	
	Physalis mild chlorosis	
	Physalis vein blotch	
	Raspberry leaf curl	
	Strawberry mild yellow edge	
	Tobacco vein distorting	
	Tobacco yellow net	
	Tobacco yellow vein assistor	
	Tomato yellow net	

Derivation of name: luteo: from Latin *luteus*, 'yellow', from yellowing symptoms shown by infected hosts

References

Duffus, J.E.: Aphids, viruses, and the yellow plague; in Harris, Maramorosch, Aphids as virus vectors, pp. 361–383 (Academic Press, New York 1977).

Iwaki, M.; Roechan, M.; Hibino, H.; Tochihara, H.; Tantera, D.M.: A persistent aphid-borne virus of soybean, Indonesian soybean dwarf virus. Plant Dis., p. 1027 (1980).

Jayasena, K.W.; Hatta, T.; Francki, R.I.B.; Randles, J.W.: Luteovirus-like particles associated with subterranean clover red leaf virus infection. J. gen. Virol. *57:* 205–209 (1981).

Namba, S.; Yamashita, S.; Doi, Y.; Yora, K.: A small spherical virus associated with the Ajinashika disease of Koshu grapevine. Ann. phytopath. Soc. Japan *45:* 70–73 (1979).

Rochow, W.F.; Duffus, J.E.: Luteoviruses and yellows diseases; in Kurstak, Handbook of plant virus infections and comparative diagnosis, pp. 147–170 (Elsevier/North Holland, Amsterdam 1981).

Rochow, W.F.; Israel, H.W.: Luteovirus (barley yellow dwarf virus) group; in Maramorosch, The atlas of insect and plant viruses, pp. 363–369 (Academic Press, New York 1977).

Yu, T.F.; Hsu, H.-K.; Pei, M.-Y.: Studies on the red-leaf disease of the foxtail millet [*Setaria italica,* (L) Beauv.]. II. Cultivated and wild hosts of millet red leaf virus. Acta phytopath. sin. *4:* 1–7 (1958).

Taxonomic status	English vernacular name	International name
Group	Tomato bushy stunt virus group	***TOMBUSVIRUS***
Type member	Tomato bushy stunt virus (TBSV) (69)	–

Main characteristics

A. Properties of the Virus Particle

Nucleic acid: One molecule of linear positive-sense ssRNA. $MW \simeq 1.5 \times 10^6$; 17% by weight of virus.
Protein: One major coat polypeptide with $MW \simeq 41 \times 10^3$.
Lipid: None reported.
Carbohydrate: None reported.
Physicochemical properties: $MW \simeq 8.9 \times 10^6$; $S_{20w} \simeq 140$; density in CsCl = 1.35 g/cm^3.
Morphology: Isometric particles with rounded outline, 30 nm in diameter. 180 protein subunits are arranged in a T=3 icosahedral lattice. In TBSV each protein subunit folds into two distinct globular domains (P and S), connected by a flexible hinge and a flexibly linked N-terminal arm. Each domain P forms one-half of a dimer-clustered surface protrusion. Domain S forms the icosahedral shell. The inward projecting N-terminal arms may have an RNA-binding function.
Antigenic properties: Good immunogens. Single precipitin line in gel diffusion tests. Serological cross-reactivity between most members.

B. Replication

1. Molecular biology: Two virus-specific dsRNA species (MW 3.2 and 1.5×10^6) have been found in tissues infected with TBSV, the smaller possibly representing the RF of a subgenomic RNA fragment.
2. Cytopathology: Cytoplasmic, compact membranous inclusions ('multivesicular bodies') are elicited by all members during early stages of infection. Virus particles are located both in cytoplasm and nuclei, often associated with the nucleolus. Virions are sometimes in crystalline arrays. Cytoplasmic accumulations of virus particles often protrude into the vacuole.

C. Biological Aspects

Host range: Wide among angiosperms.
Transmission: Readily transmitted by mechanical inoculation. Acquisition through soil is possible.

Other members

Artichoke mottled crinkle
Carnation Italian ringspot
Cymbidium ringspot (178)
Eggplant mottled crinkle
Pelargonium leaf curl
Petunia asteroid mosaic

Possible members

Saguaro cactus (148)
Turnip crinkle (109)

Derivation of name

tombus: sigla from *tom*ato *bu*shy *s*tunt

References

Harrison, S.C.; Olsen, A.J.; Schutt, C.E.; Winkler, F.K.; Bricogne, G.: Tomato bushy stunt virus at 2.9 Å resolution. Nature, Lond. *276:* 368–373 (1978).

Henriques, M.-I.C.; Morris, T.J.: Evidence for different replicative strategies in the plant tombusviruses. Virology *99:* 66–74 (1979).

Hollings, M.; Stone, O.M.: Serological and immunoelectrophoretic relationship among viruses in the tombusvirus group. Ann. appl. Biol. *80:* 37–48 (1975).

Hollings, M.; Stone, O.M.; Barton, R.J.: Pathology, soil transmission and characterization of cymbidium ringspot, a virus from cymbidium orchids and white clover *(Trifolium repens)*. Ann. appl. Biol. *85:* 233–248 (1977).

Makkouk, K.M.; Koenig, R.; Lesemann, D.E.: Characterization of a tombusvirus isolated from eggplant. Phytopathology *71:* (in press, 1981).

Martelli, G.P.: Tombusviruses; in Kurstak, Plant virus infections and comparative diagnosis, pp. 61–90 (Elsevier/North Holland, Amsterdam 1981).

Martelli, G.P.; Russo, M.; Quacquarelli, A.: Tombusvirus (tomato bushy stunt virus) group; in Maramorosch, The atlas of insect and plant viruses, pp. 257–279 (Academic Press, New York 1977).

Ziegler, A.; Harrison, S.C.; Leberman, R.: The minor proteins in tomato bushy stunt and turnip crinkle viruses. Virology *59:* 509–515 (1974).

Taxonomic status	*English vernacular name*	*International name*
Group	Southern bean mosaic virus group	***SOBEMOVIRUS***
Type member	Southern bean mosaic virus (SBMV) (57)	–
Main characteristics	A. Properties of the Virus Particle Nucleic acid: One molecule of positive-sense ssRNA; MW$\simeq$1.4 × 10^6; low-molecular-weight protein which is essential for infectivity of RNA (SBMV, turnip rosette virus) is associated with 5′-end; 3′-end does not contain poly(A) or a tRNA-like structure. Protein: One coat polypeptide, MW$\simeq$30 × 10^3. Lipid: None. Carbohydrate: None. Physicochemical properties: MW$\simeq$6.6 × 10^6; density $\simeq$1.36 g/cm^3 in CsCl (but virus forms two or more bands in Cs_2SO_4); $S_{20w}$$\simeq$115; swelling of particles due to EDTA. Chelation of divalent cations is reversible by addition of cations or by decreasing pH. Morphology: Particles $\simeq$30 nm diameter with 180 subunits in a T = 3 icosahedral structure stabilized by divalent cations. Each protein subunit has two domains. One forms parts of the icosahedral shell about 3.5 nm thick and the other forms a partially ordered 'arm' in the interior of the virus. Antigenic properties: Efficient immunogens. Single precipitin line in gel diffusion tests. Serological relationships between strains but not between other members of the group. B. Replication In vitro translation of 1.4 × 10^6 RNA of SBMV in wheat embryo system or of turnip rosette virus RNA in rabbit reticulocyte lysates yields three proteins but no coat protein; the latter is translated only from 0.3–0.4 × 10^6 RNA which is present in RNA isolated from purified virus, suggesting the possibility that the coat protein message is subgenomic RNA. In situ, virions are located in both nuclei and cytoplasm; those in cytoplasm sometimes form crystalline arrays. C. Biological Aspects Host range: Each virus has relatively narrow host range. Transmission: Seed transmission in several host plants. Transmitted by beetles. Readily transmitted mechanically.	
Other member	Turnip rosette virus (125)	
Possible members	Blueberry shoestring (204) Cocksfoot mottle (23) Rice yellow mottle (149) Sowbane mosaic (64)	
Derivation of name	sobemo: sigla derived from the name of the type member *so*uthern *be*an *mo*saic	

References

Abad-Zapatero, C.; Abdel-Meguid, S.S.; Johnson, J.E.; Leslie, A.G.W.; Rayment, I.; Rossmann, M.G.; Suck, D.; Tsukihara, T.: Structure of Southern bean mosaic virus at 2.8 Å resolution. Nature, Lond. *286:* 33–39 (1980).

Ghosh, A.; Dasgupta, R.; Salerno-Rife, T.; Rutgers, T.; Kaesberg, P.: Southern bean mosaic viral RNA has a 5′-linked protein but lacks 3′ terminal poly(A). Nucl. Acids Res. *7:* 2137–2146 (1979).

Hull, R.: The grouping of small spherical plant viruses with single RNA components. J. gen. Virol. *36:* 289–295 (1977a).

Hull, R.: The stabilization of the particles of turnip rosette virus and of other members of the southern bean mosaic virus group. Virology *79:* 58–66 (1977b).

Hull, R.; Morris-Krsinich, B.A.M.: Protein covalently linked to the RNA of turnip rosette virus. John Innes Inst. 17th Annu. Rep., pp.98–99 (1979).

Morris-Krsinich, B.A.M.: *In vitro* translation of turnip rosette virus RNA in rabbit reticulocyte lysates. John Innes Inst. 17th Annu. Rep., p. 100 (1979).

Salerno-Rife, T.; Rutgers, T.; Kaesberg, P.: Translation of southern bean mosaic virus RNA in wheat embryo and rabbit reticulocyte extracts. J. Virol. *34:* 51–58 (1980).

Taxonomic status	*English vernacular name*	*International name*
Group	Tobacco necrosis virus group	–
Type member	Tobacco necrosis virus (TNV) (A strain) (14)	–
Main characteristics	A. Properties of the Virus Particle Nucleic acid: One molecule of linear positive-sense ssRNA with MW$\simeq$1.3–1.6 $\times 10^6$. 5′ terminus has the sequence ppApGpUp... Protein: Single polypeptide with MW$\simeq$22.6 $\times 10^3$. 180 protein subunits arranged in a T = 3 icosahedral lattice. Lipid: None. Carbohydrate: None. Physicochemical properties: MW$\simeq$7.6 $\times 10^6$; $S_{20w}\simeq$118; density in CsCl$\simeq$1.40 g/cm^3. Morphology: Polyhedral particles $\simeq$28 nm diameter. Antigenic properties: Moderately immunogenic. Single precipitin line in gel diffusion tests. B. Replication A virus-induced RNA-dependent polymerase occurs within infected plants. Crystal-like aggregates of virus particles often seen in cytoplasm of infected cells. C. Biological Aspects Host range: Wide among angiosperms. Transmission: Transmitted naturally by the chytrid fungus *Olpidium,* and experimentally by mechanical inoculation of sap.	
Possible member	Cucumber necrosis (82)	

References

Kassanis, B.: Tobacco necrosis virus group; in Maramorosch, The atlas of insect and plant viruses, pp. 281–285 (Academic Press, New York 1977).

Kassanis, B.; Vince, D.A.; Woods, R.D.: Light and electron microscopy of cells infected with tobacco necrosis and satellite viruses. J. gen. Virol. *7:* 143–151 (1970).

Stussi-Garaud, C.; Lemius, J.; Fraenkel-Conrat, H.: RNA polymerase from tobacco necrosis virus-infected and uninfected tobacco. Virology *81:* 224–236 (1977).

Tremaine, J.H.: Purification and properties of cucumber necrosis virus and a smaller top component. Virology *48:* 582–590 (1972).

Uyemoto, J.K.: Tobacco necrosis and satellite viruses; in Kurstak, Handbook of plant virus infections and comparative diagnosis, pp. 123–146 (Elsevier/North Holland, Amsterdam 1981).

Taxonomic status	*English vernacular name*	*International name*
Group	Beet yellows virus group	***CLOSTEROVIRUS***
Type member	Sugar beet yellows virus (SBYV) (13)	–
Main characteristics	A. Properties of the Virus Particle Nucleic acid: One molecule of linear positive-sense ssRNA, MW$\simeq$2.2–4.7 $\times 10^6$; about 5% by weight of virus particle. Protein: One coat polypeptide, MW$\simeq$23–27 $\times 10^3$. Lipid: None reported. Carbohydrate: None reported. Physicochemical properties: $S_{20w}\simeq$96–130; density in CsCl$\simeq$1.30–1.34 g/cm^3. Particles unstable in high salt concentrations. Morphology: Very flexuous rods 600–2,000 nm long and 12 nm wide. Helical symmetry with pitch$\simeq$3.4–3.7 nm. Antigenic properties: Moderately immunogenic; serological relationships between some members. B. Replication Particles often aggregate in cross-banded masses in phloem cells. No data on molecular biology. C. Biological Aspects Host range: Moderately wide for individual viruses. Transmission: Experimentally transmissible with difficulty by mechanical inoculation; some members transmitted by aphids in a semi-persistent manner.	
Other members	Beet yellow stunt (207) Burdock yellows Carnation necrotic fleck (136) Carrot yellow leaf Citrus tristeza (33) Clover yellows *Festuca* necrosis Grapevine stem-pitting associated Lilac chlorotic leafspot (202) Wheat yellow leaf (157)	
Possible members	Apple chlorotic leafspot (30) Apple stem grooving (31) *Heracleum* latent (228) Potato T (187)	
Derivation of name	clostero: from Greek *kloster,* 'spindle, thread', from appearance of very long rods	

References

Bar-Joseph, M.; Garnsey, S.M.; Gonsalves, D.: The closteroviruses: a distinct group of elongated plant viruses. Adv. Virus Res. *25:* 93–168 (1979).

Boccardo, G.; D'Aquitio, M.: The protein and nucleic acid of a closterovirus isolated from a grapevine with stem-pitting symptoms. J. gen. Virol. *53:* 179–182 (1981).
Lister, R.M.; Bar-Joseph, M.: Closteroviruses; in Kurstak, Handbook of plant virus infections and comparative diagnosis, pp. 809–844 (Elsevier/North Holland, Amsterdam 1981).

Taxonomic status	*English vernacular name*	*International name*
Group	Carnation latent virus group	***CARLAVIRUS***
Type member	Carnation latent virus (CLV) (61)	–
Main characteristics	A. Properties of the Virus Particle Nucleic acid: One molecule of linear positive-sense ssRNA; MW$\simeq 2.7 \times 10^6$; $\simeq$6% by weight of virus. Protein: One coat polypeptide, MW$\simeq 32 \times 10^3$. Lipid: Probably none. Carbohydrate: Not known. Physicochemical properties: $S_{20w} \simeq 160$. Buoyant density in CsCl$\simeq$1.3 g/cm^3. Morphology: Slightly flexuous rods 600–700 nm long, $\simeq$13 nm diameter. Helical symmetry with pitch $\simeq$3.4 nm. Antigenic properties: Serological relationships among members range from medium to not detectable. B. Replication Molecular biological aspects not known. With several members, aggregates of particles in paracrystalline arrays, sometimes banded, and inclusions containing masses of endoplasmic reticulum interspersed with ribosomes and unaggregated virus particles are seen in the cytoplasm. C. Biological Aspects Host range: Individual viruses have rather narrow host ranges. Transmission: Experimentally by mechanical inoculation. Often by aphids in a non-persistent manner.	
Other members	Cactus 2 *Chrysanthemum* B (110) Cowpea mild mottle (140) Elderberry carla *Helenium* S Hop latent Hop mosaic (241) Lilac mottle Lily symptomless (96) *Lonicera* latent (=Honeysuckle latent?) Mulberry latent Muskmelon vein necrosis *Narcissus* latent (170) *Nerine* latent *Passiflora* latent Pea streak (=Alfalfa latent?) (112) Pepino latent Poplar mosaic (75) Potato M (87) Potato S (60)	

Taxonomic status	*English vernacular name*	*International name*
	Red clover vein mosaic (22) Shallot latent	
Possible members	Artichoke latent Caper vein banding *Cassia* mild mosaic Chicory blotch Cole latent *Cynodon* mosaic Eggplant mild mottle *Fuchsia* latent Garlic mosaic Groundnut crinkle *Gynura* latent (strain of *Chrysanthemum* B ?) Nasturtium mosaic	
Derivation of name	carla: sigla from *car*nation *la*tent	

References

Adams, A.N.; Barbara, D.J.: Host range, purification and some properties of hop mosaic virus. Ann. appl. Biol. *96:* 201–208 (1980).

Bos, L.; Huttinga, H.; Maat, D.Z.: Shallot latent virus, a new carlavirus. Neth. J. Plant. Pathol. *84:* 227–237 (1978).

Brunt, A.A.; Phillips, S.; Thomas, B.J.: Honeysuckle latent virus, a carlavirus infecting *Lonicera periclymenum* and *L. japonica (Caprifoliaceae)*. Acta horticult. *110:* 205–209 (1980).

Cadilhac, B.; Quiot, J.B.; Marrou, J.; Leroux, J.P.: Mise en évidence au microscope électronique de deux virus différents infectant l'ail (*Allium sativum* L.) et l'échalote (*Allium cepa* L. var. Ascolanicum). Ann. Phytopathol. *8:* 65–72 (1976).

Dubern, J.; Dollet, M.: Groundnut crinkle, a new virus disease observed in Ivory Coast. Phytopath. Z. *95:* 279–283 (1979).

Gumpf, D.J.; Osman, F.M.; Weathers, L.G.: Purification and some properties of a latent virus in Gynura. Plant Dis. Reptr *61:* 325–327 (1977).

Hampton, R.O.: Evidence suggesting identity between alfalfa latent and pea streak viruses. Phytopathology *71:* 223 (1981).

Johns, L.J.; Stace-Smith, R.; Kadota, D.Y.: Occurrence of a rod-shaped virus in Fuchsia cultivars. Acta horticult. *110:* 195–203 (1980).

Khalil, J.A.; Nelson, M.R.; Wheeler, R.E.: Eggplant mild mottle virus – a new carlavirus. Phytopath. News *12:* 169 (1978).

Lin, M.T.; Kitajima, E.W.; Cupertino, F.P.; Costa, C.L.: Properties of possible carlavirus isolated from a Cerrado native plant *Cassia sylvestris*. Plant Dis. Reptr *63:* 501–505 (1979).

Majorana, G.: La reticolatura fogliare del Capparo: una malattia assocata ad un virus del gruppo S della patata. Phytopath. Mediterranea *9:* 106–110 (1979).

Majorana, G.; Rana, G.L.: A latent virus of artichoke belonging to the potato virus S group. Phytopath. Mediterranea *9:* 200–202 (1970).

Thomas, W.; Mohamed, N.A.; Fry, M.E.: Properties of a carlavirus causing latent infection of pepino *(Solanum muricatum)*. Ann. appl. Biol. *95:* 191–196 (1980).

van der Meer, F.A.; Maat, D.Z.; Vink, J.: Lonicera latent virus, a new carlavirus serologically related to poplar mosaic virus: some properties and inactivation *in vivo* by heat treatment. Neth. J. Plant Pathol. *86:* 69–78 (1980).

van Lent, J.W.M.; Wit, A.J.; Dijkstra, J.: Characterization of a carlavirus in elderberry (*Sambucus* spp.). Neth. J. Plant Pathol. *86:* 117–134 (1980).

Wetter, C.; Milne, R.G.: Carlaviruses; in Kurstak, Handbook of plant virus infections and comparative diagnosis, pp. 695–730 (Elsevier/North Holland, Amsterdam 1981).

Taxonomic status	*English vernacular name*	*International name*
Group	Potato virus Y group (245)	***POTYVIRUS***
Type member	Potato virus Y (PVY)	–

Main characteristics

A. Properties of the Virus Particle

Nucleic acid: One molecule of linear positive-sense ssRNA. MW = $3.0–3.5 \times 10^6$; $\simeq 5\%$ by weight of particle. RNA molecules of some viruses have poly(A) tracts at their 3′ ends.
Protein: One coat polypeptide, MW = $32–36 \times 10^3$.
Lipid: None reported.
Carbohydrate: None reported.
Physicochemical properties: $S_{20w} = 150–160$; density in CsCl = 1.31 g/cm^3.
Morphology: Flexuous filaments 680–900 nm long and 11 nm wide, with helical symmetry and pitch = 3.4 nm.
Antigenic properties: Moderately immunogenic; serological relationships between some members.

B. Replication

Characteristic cylindrical or conical inclusions, appearing as pinwheels when seen in transverse section, are induced in the cytoplasm; protein of inclusions (MW = 70×10^3) serologically unrelated to virus coat protein but specified by viral genome. Some members also induce nuclear inclusions. RNA from some members has been translated in vitro into proteins of MW corresponding to more than 90% of the genome coding potential.

C. Biological Aspects

Host range: Narrow for individual viruses.
Transmission: Transmissible experimentally by mechanical inoculation; transmitted by aphids in a non-persistent manner. Others, very tentatively included as possible members of the group, are transmissible by whiteflies, mites or fungi.

Other members

Amaranthus leaf mottle
Bean common mosaic (73)
Bean yellow mosaic (= pea mosaic) (40)
Bearded iris mosaic (147)
Beet mosaic (53)
Bidens mottle (161)
Blackeye cowpea mosaic
Carnation vein mottle (78)
Carrot thin leaf (218)
Celery mosaic (50)
Clover yellow vein (= pea necrosis) (131)
Cocksfoot streak (59)
Colombian datura
Commelina mosaic
Cowpea aphid-borne mosaic (= Azuki bean mosaic) (134)
Dasheen mosaic (191)
Datura shoestring
Guinea grass mosaic (190)
Henbane mosaic (95)

Taxonomic status	*English vernacular name*	*International name*
	Hippeastrum mosaic (117)	
	Iris mild mosaic (116)	
	Iris severe mosaic	
	Leek yellow stripe (240)	
	Lettuce mosaic (9)	
	Narcissus degeneration	
	Nothoscordum mosaic	
	Onion yellow dwarf (159)	
	Papaya ringspot (84)	
	Parsnip mosaic (91)	
	Passionfruit woodiness (122)	
	Pea seed-borne mosaic (146)	
	Peanut mottle (141)	
	Pepper mottle	
	Pepper severe mosaic	
	Pepper veinal mottle (104)	
	Plum pox (70)	
	Pokeweed mosaic (97)	
	Potato A (54)	
	Soybean mosaic (93)	
	Sugarcane mosaic (=maize dwarf mosaic) (88)	
	Tamarillo mosaic	
	Tobacco etch (55)	
	Tulip breaking (71)	
	Turnip mosaic (8)	
	Watermelon mosaic 1	
	Watermelon mosaic 2	
	Wisteria vein mosaic	
Possible members	*Aphid-borne:*	
	Anthoxanthum mosaic	
	Aquilegia	
	Araujia mosaic	
	Artichoke latent	
	Bidens mosaic	
	Bryonia mottle	
	Canavalia maritima mosaic	
	Carrot mosaic	
	Celery yellow mosaic	
	Clover (Croatian)	
	Crinum	
	Daphne Y	
	Datura 437	
	Datura mosaic	
	Dendrobium mosaic	
	Desmodium mosaic	
	Dioscorea green banding (=yam mosaic)	
	Dioscorea trifida	

Taxonomic status	English vernacular name	International name
	Dock mottling mosaic	
	Euphorbia ringspot	
	Fern	
	Freesia mosaic	
	Garlic yellow streak	
	Gloriosa stripe mosaic	
	Groundnut eyespot	
	Guar symptomless	
	Helenium virus Y	
	Holcus streak	
	Hyacinth mosaic	
	Iris fulva mosaic	
	Isachne mosaic	
	Kennedya Y	
	Maclura mosaic (239)	
	Malva vein clearing	
	Marigold mottle	
	Mungbean mosaic	
	Mungbean mottle	
	Narcissus late season yellows (=jonquil mild mosaic)	
	Narcissus yellow stripe (76)	
	Nerine	
	Ornithogalum mosaic	
	Palm mosaic	
	Passionfruit ringspot	
	Primula mosaic	
	Reed canary mosaic	
	Statice Y	
	Sweet potato A	
	Sweet potato russet crack	
	Teasel mosaic	
	Tobacco vein mottling	
	Tomato (Peru) mosaic	
	Tradescantia/Zebrina	
	Wheat spindle streak (167)	
	Wheat streak	
	White bryony mosaic	
	Wild potato mosaic	
	Fungal-borne:	
	Barley yellow mosaic (143)	
	Oat mosaic (145)	
	Rice necrosis mosaic (172)	
	Wheat spindle streak mosaic	
	Wheat yellow mosaic	
	Mite-borne:	
	Agropyron mosaic (118)	
	Oat necrotic mottle (169)	
	Ryegrass mosaic (86)	

Taxonomic status	*English vernacular name*	*International name*
	Spartina mottle	
	Wheat streak mosaic	
	Whitefly-borne:	
	Sweet potato mild mottle (162)	

Derivation of name	poty: sigla from *pot*ato *Y*

References

Bos, L.; Huijberts, N.; Huttinga, H.; Maat, D.Z.: Leek yellow stripe virus and its relationships to onion yellow dwarf virus: characterization, ecology and possible control. Neth. J. Plant Pathol. *84:* 185–204 (1978).

Brunt, A.A.: Narcissus yellow stripe virus (NYSV). Glasshouse Crops Res. Inst. Annu. Rep., pp. 133–134 (1969).

Brunt, A.A.; Phillips, S.: *Narcissus narcissus* spp. Glasshouse Crops Res. Inst. Annu. Rep., p. 130 (1977).

Charudattan, R.; Zettler, F.W.; Cordo, H.A.; Christie, R.G.: Partial characterization of a potyvirus infecting the milkweed vine, *Morrenia odorata*. Phytopathology *70:* 909–913 (1980).

Dougherty, W.G.; Hiebert, E.: Translation of potyvirus RNA in rabbit reticulocyte lysate: cell-free translation strategy and a genetic map of the potyviral genome. Virology *104:* 183–194 (1980).

Edwardson, J.R.: Some properties of the potato virus Y group. Florida agric. exp. Station Monogr. Ser., vol. 4 (1974a).

Edwardson, J.R.: Host ranges of viruses in the PVY group. Florida agric. exp. Station Monogr. Ser., vol. 5 (1974b).

Hari, V.; Siegel, A.; Ruzek, C.; Timberlake, W.E.: The RNA of tobacco etch virus contains poly(A). Virology *92:* 568–571 (1979).

Hollings, M.; Brunt, A.A.: Potyviruses; in Kurstak, Plant virus infections and comparative diagnosis, pp. 731–807 (Elsevier/North Holland, Amsterdam 1981).

Horvat, F.; Verhoyen, M.: Inclusions in mesophyll cells induced by a virus causing chlorotic streaks on leaves of *Allium porrum* L. Phytopath. Z. *83:* 328–340 (1975).

Kitajima, E.W.; Costa, A.S.: The fine structure of the intranuclear, fibrous inclusions associated with the infection by celery yellow mosaic virus. Fitopatol. bras. *3:* 287–293 (1978).

Kuschki, G.H.; Koenig, R.; Duvel, D.; Kuhne, H.: Helenium virus S and Y – two new viruses from commercially grown *Helenium* hybrids. Phytopathology *68:* 1407–1411 (1978).

Lockhart, B.E.L.; Betzold, J.A.: Some properties of a potyvirus occurring in *Tradescantia* and *Zebrina* spp. Acta horticult. *110:* 55–57 (1980).

Moyer, J.W.; Kennedy, G.G.: Purification and properties of sweet potato feathery mottle virus. Phytopathology *68:* 998–1004 (1978).

Naqvi, Q.A.; Hadi, S.; Mahmood, K.: Marigold mottle virus in Aligarh, India. Plant Dis. *65:* 271–275 (1981).

Nienhaus, F.: The feathery mottle syndrome of sweet-potato *(Ipomoea batatas)* in Togo. Z. Pflanzenkrankh. Pflanzenschutz *87:* 185–189 (1980).

Rodriguez, R.L.; Bird, J.; Monllar, A.C.; Waterworth, H.E.; Kimura, M.; Maramorosch, K.: The mosaic virus of *Canavalia maritima* (Bay-bean) in Puerto Rico; in Bird, Maramorosch, Tropical diseases of legumes, pp. 91–101 (Academic Press, New York 1975).

Thouvenel, J.C.; Fanquet, C.: Yam mosaic, a new potyvirus infecting *Dioscorea cayenensis* in the Ivory Coast. Ann. appl. Biol. *93:* 279–283 (1979).

Tomlinson, J.A.; Carter, A.L.: Virus diseases of umbelliferous plants. Natn. Vegetable Res. Station Annu. Rep. *20:* 108 (1969).

Toriyama, S.; Yora, K.: Virus diseases of wild grasses and cereal crops in Japan (University of Tokyo Press, Tokyo 1972).

Webb, M.J.W.: Electron microscopy. Natn. Vegetable Res. Station Annu. Rep. *21:* 113–114 (1970).

Taxonomic status	*English vernacular name*	*International name*
Group	Potato virus X group	***POTEXVIRUS*** (200)
Type member	Potato virus X (PVX) (4)	–

Main characteristics

A. Properties of the Virus Particle

Nucleic acid: One molecule of linear positive-sense ssRNA; MW$\simeq$2.1 × 10^6; 5% by weight of the particle. 5′ terminus has sequence $m^7G^{5'}$pppGpA... No poly(A) at 3′ terminus but RNA contains high A content ($\simeq$30%).

Protein: One coat polypeptide, MW$\simeq$18–23 × 10^3. In some viruses, protein can become partially degraded by enzymes in plant sap.

Lipid: None reported.

Carbohydrate: None reported.

Physicochemical properties: MW$\simeq$35 × 10^6; S_{20w} = 115–130; density in CsCl$\simeq$1.31 g/cm^3; particles stable.

Morphology: Flexuous rods 470–580 nm long and 13 nm wide, with helical symmetry and pitch$\simeq$3.4 nm.

Antigenic properties: Efficient immunogens; serological relationship between some members.

B. Replication

Molecular biology: Intact genomic RNA is translated into high-molecular-weight proteins, but apparently not into viral coat protein.

Cytopathology: Fibrous cytoplasmic inclusions composed of virus particles, often banded; some members induce nuclear inclusions of different composition.

C. Biological Aspects

Host range: Narrow for individual viruses.

Transmission: Readily transmitted mechanically, experimentally, and by contact between plants. No known vectors.

Other members

Boussingaultia mosaic
Cactus X (58)
Cassava common mosaic
Clover yellow mosaic (111)
Commelina X
Cymbidium mosaic (27)
Foxtail mosaic
Hydrangea ringspot (114)
Lily X
Narcissus mosaic (45)
Nerine X
Papaya mosaic (56)
Pepino mosaic
Plantago severe mottle
Plantago X
Viola mottle
White clover mosaic (41)

Taxonomic status	*English vernacular name*	*International name*
Possible members	Artichoke curly dwarf	
	Bamboo mosaic	
	Barley B-1	
	Boletus	
	Centrosema mosaic	
	Daphne X (195)	
	Dioscorea latent	
	Hippeastrum latent	
	Lily	
	Malva veinal necrosis	
	Nandina mosaic	
	Negro coffee mosaic	
	Parsley 5	
	Parsnip 3	
	Potato aucuba mosaic (98)	
	Rhododendron necrotic ringspot	
	Rhubarb 1	
	Wineberry latent	
	Zygocactus	

Derivation of name	potex: sigla from *pot*ato *X*

References

Beczner, L.; Vassányi, R.: Identification of a new potexvirus isolated from *Boussingaultia cordifolia* and *B. gracilis* f. pseudo-baselloides. Tag.-Ber. Akad. Landwirtsch.-Wiss. DDR *184:* 65–75 (1980).

Hammond, J.; Hull, R.: Plantain virus X: a new potexvirus from *Plantago lanceolata*. J. gen. Virol. *54:* 75–90 (1981).

Purcifull, D.E.; Edwardson, J.R.: Potexviruses; in Kurstak, Handbook of plant virus infections and comparative diagnosis, pp. 627–693 (Elsevier/North Holland, Amsterdam 1981).

Rowhani, A.; Peterson, J.F.: Characterization of a flexuous rod-shaped virus from Plantago. Can. J. Plant Pathol. *2:* 12–18 (1980).

Short, M.N.: Foxtail mosaic virus, a new member of the potexvirus group. Abstr. 5th Int. Congr. Virol., Strasbourg, p. 239 (l'Imprimerie Centrale Commerciale, Paris 1981).

Stone, O.M.: Two new potexviruses from monocotyledons. Acta horticult. *110:* 59–70 (1980).

Wodnar-Filipowicz, A.; Skrzeczkowski, L.J.; Filipowicz, W.: Translation of potato virus X RNA into high molecular weight proteins. FEBS Lett. *109:* 151–155 (1980).

Taxonomic status	*English vernacular name*	*International name*
Group	Tobacco mosaic virus group	***TOBAMOVIRUS*** (184)
Type member	Tobacco mosaic virus (TMV) (common or U1 strain) (151)	–
Main characteristics	A. Properties of the Virus Particle Nucleic acid: One molecule of linear positive-sense ssRNA; MW$\simeq 2\times 10^6$. 5′ terminus has the sequence $m^7G^{5\prime}ppp^{5\prime}Gp$. 3′ terminus has a tRNA-like structure which accepts histidine. Protein: One coat polypeptide, MW$\simeq$17–18$\times 10^3$. Lipid: None. Carbohydrate: None. Physicochemical properties: MW$\simeq 40\times 10^6$; $S_{20w}\simeq 194$; density in CsCl$\simeq$1.325 g/cm^3; particles very stable. Morphology: Elongated rigid particles about 18 nm diameter and 300 nm long, helically symmetrical with pitch$\simeq$2.3 nm. Antigenic properties: Efficient immunogens. B. Replication Virus replicates in the cytoplasm, inducing characteristic viroplasms; virus particles often form large crystalline arrays, visible by light microscopy. A virus-induced polymerase is detected in infected tissues; RNA replicates via an RF or RI. Coat protein is synthesized from a small monocistronic mRNA (whose base sequence is also on the viral RNA near the 3′ end); the mRNA is encapsidated in some virus strains. Three other virus-specific proteins (MW$\simeq$165, 100 and 30$\times 10^3$) are transcribed from full-length viral RNA. C. Biological Aspects Host range: Most members have moderate host range. Transmission: Readily transmitted by mechanical inoculation. Some members transmitted by seed.	
Other members	Cucumber green mottle mosaic (154) Cucumber 4 Frangipani mosaic (196) *Odontoglossum* ringspot (155) Ribgrass mosaic (152) Sammon's *Opuntia* Sunnhemp mosaic (153) Tomato mosaic (156) U2-tobacco mosaic	
Possible members	Beet necrotic yellow vein mosaic (144) *Chara australis* *Nicotiana velutina* mosaic (189) Peanut clump (235) Potato mop top (138) Soil-borne wheat mosaic (77)	

Taxonomic status	*English vernacular name*	*International name*
Derivation of name	tobamo: sigla from *toba*cco *mo*saic	

References

Hirth, L.; Richards, K.E.: Tobacco mosaic virus: model for structure and function of a simple virus. Adv. Virus Res. *26:* 145–199 (1981).

Shikata, E.: Tobamovirus (tobacco mosaic virus) group; in Maramorosch, The atlas of insect and plant viruses, pp. 237–255 (Academic Press, New York 1977).

Van Regenmortel, M.H.V.: Tobamoviruses; in Kurstak, Handbook of plant virus infections and comparative diagnosis, pp. 541–564 (Elsevier/North Holland, Amsterdam 1981).

Taxonomic status	*English vernacular name*	*International name*
Group	Carnation ringspot virus group	***DIANTHOVIRUS***
Type member	Carnation ringspot virus (CRSV) (21)	–
Main characteristics	A. Properties of the Virus Particle Nucleic acid: Two molecules of positive-sense ssRNA; MW$\simeq$1.5 and 0.5×10^6. The larger RNA contains the coat protein cistron. Protein: One coat polypeptide, with MW$\simeq 40 \times 10^3$ Lipid: None reported. Carbohydrate: None reported. Physicochemical properties: MW$\simeq 7.1 \times 10^6$; density in CsCl$\simeq$1.37 g/cm^3; $S_{20w} \simeq 135$; particles dissociate in EDTA. Morphology: Polyhedral particles 31–34 nm diameter. The arrangement of the two RNA species within particles has not been established. Antigenic properties: Efficient immunogens. Single precipitin line in gel diffusion tests. B. Replication Particles located in the cytoplasm, scattered and clustered; patches of densely stained, amorphous material also seen in cytoplasm of some cells. C. Biological Aspects Host range: Each virus has a wide host range. Transmission: Readily transmissible experimentally by mechanical inoculation. Transmitted through soil.	
Other members	Red clover necrotic mosaic (181) Sweet clover necrotic mosaic	
Derivation of name	From *Dianthus*, the generic name of carnation	

References

Dodds, J.A.; Tremaine, J.H.; Ronald, W.P.: Some properties of carnation ringspot virus single- and double-stranded ribonucleic acid. Virology *83:* 322–328 (1977).

Gould, A.R.; Francki, R.I.B.; Hatta, T.; Hollings, M.: The bipartite genome of red clover necrotic mosaic virus. Virology *108:* 499–506 (1981).

Hiruki, C.; Okuno, T.; Rao, T.; Rao, D.V.; Chen, M.H.: A new bipartite genome virus, sweet clover necrotic mosaic virus occurring in Alberta. Abstr. 5th Int. Congr. Virol., Strasbourg, p. 235 (l'Imprimerie Centrale Commerciale, Paris 1981).

Taxonomic status	*English vernacular name*	*International name*
Group	Cowpea mosaic virus group	***COMOVIRUS*** (199)
Type member	Cowpea mosaic virus (CPMV) (SB isolate) (47, 197)	–

Main characteristics A. Properties of the Virus Particle

Nucleic acid: Two species of linear positive-sense ssRNA with MW$\simeq 2.4 \times 10^6$ (RNA-1) and 1.4×10^6 (RNA-2). The two RNA molecules each have a high content of A+U but have little base sequence homology. Each molecule has a poly(A) tract $\simeq$120 residues long at its 3′ end and a small polypeptide (MW$\simeq$5,000) covalently linked to its 5′ end; enzymatic degradation of this polypeptide does not diminish the infectivity of the RNA. The cistrons for the genome-linked protein, and for a proteolytic protein that cleaves RNA-2 products, are on RNA-1.

Protein: Two coat polypeptides, MWs$\simeq$22 and 42×10^3.

The smaller, and in some members both, polypeptide may become partially degraded by proteolytic cleavage in vivo and in vitro.

Lipid: None reported.

Carbohydrate: The coat proteins may be glycosylated.

Physicochemical properties: Particles usually very stable and sediment as three components, T, M and B, respectively containing$\simeq$0, 25 and 37% RNA by weight with $S_{20w} \simeq 58$, 98 and 118, and MWs$\simeq$3.8, 5.2, and 6.2×10^6. Diffusion coefficients of all three particles are about 1.30×10^{-7} cm^2/s. Buoyant densities in CsCl$\simeq$1.29 (T), 1.41 (M) and 1.44–1.46 (B) g/cm^3. Partial proteolytic degradation of the smaller coat protein results in particles with increased electrophoretic mobility.

Morphology: All three sedimenting components possess isometric particles$\simeq$28 nm in diameter. The shell consists of 60 subunits of each of the two structural proteins assembled in a T=1 icosahedral structure. There are probably 12 pentamers of the larger protein at the 5-fold vertices and 20 trimers of the smaller protein at the positions of 3-fold symmetry. M particles contain a single molecule of RNA-2, and B particles a single molecule of RNA-1.

Antigenic properties: Good immunogens. All members are serologically inter-related, often distantly.

B. Replication

Unfractionated RNA is highly infective but neither RNA species alone can infect plants. RNA-1 can replicate in protoplasts but, in the absence of RNA-2 (which carries the coat protein cistron), no virus particles are produced. RNA-1 presumably carries information for the polymerase function. Membranous vesicular structures appear as characteristic inclusion bodies in the cytoplasm, usually adjacent to the nuclei. They contain virus-induced RNA-dependent RNA polymerase, two dsRNA species corresponding to each of the particle RNA molecules, and complementary RNA. Inhibitor studies indicate that the CPMV proteins are synthesized in the cytoplasm on 80S ribosomes. Both RNA species are translated in vitro into large polypeptides approaching in size the theoretical coding capacity; in vivo these 'polyproteins' are cleaved to form the functional proteins. Newly formed virus particles accumulate in the cytoplasm, sometimes in crystalline arrays but not in association with any cell organelle.

C. Biological Aspects

Host range: Individual members have narrow host ranges. Mosaic and mottle symptoms are characteristic.

Taxonomic status	*English vernacular name*	*International name*
	Transmission: Readily transmissible experimentally by mechanical inoculation. Some are seed-transmitted. Some members transmitted by beetles, especially Chrysomelidae. Beetles retain ability to transmit virus for days or weeks.	
Other members	Andean potato mottle (203) Bean pod mottle (108) Bean rugose mosaic Broad bean stain (29) Broad bean true mosaic (20) Cowpea severe mosaic (209) *Glycine* mosaic Quail pea mosaic (238) Radish mosaic (121) Red clover mottle (74) Squash mosaic (43)	
Possible member	Broad bean wilt	
Derivation of name	como: sigla from *co*wpea *mo*saic	

References

Bowyer, J.W.; Dale, J.L.; Behncken, G.M.: Glycine mosaic virus: a comovirus from Australian native glycine species. Ann. appl. Biol. *95:* 385–390 (1980).

Bruening, G.: Plant covirus systems: two-component systems; in Fraenkel-Conrat, Wagner, Comprehensive virology, vol. 11, pp. 55–141 (Plenum, New York 1977).

Daubert, S.D.; Bruening, G.: Genome-associated proteins of comoviruses. Virology *98:* 246–250 (1979).

Davies, J.W.; Aalbers, A.M.J.; Stuik, E.J.; van Kammen, A.: Translation of cowpea mosaic virus RNA in a cell-free extract from wheat germ. FEBS Lett. *77:* 265–269 (1977).

Fulton, J.P.; Scott, H.A.: A serogrouping concept for legume comoviruses. Phytopathology *69:* 305–306 (1979).

Goldbach, R.; Rezelman, G.; van Kammen, A.: Independent replication and expression of B-component RNA of cowpea mosaic virus. Nature, Lond. *286:* 297–300 (1980).

Murant, A.F.; Taylor, M.; Duncan, G.H.; Raschké, J.H.: Improved estimates of molecular weight of plant virus RNA by agarose gel electrophoresis and electron microscopy after denaturation with glyoxal. J. gen. Virol. *53:* 321–332 (1981).

Pelham, H.R.B.: Synthesis and proteolytic processing of cowpea mosaic virus proteins in reticulocyte systems. Virology *96:* 463–477 (1979).

Rottier, P.J.M.; Rezelman, G.; van Kammen, A.: Protein synthesis in cowpea mosaic virus-infected cowpea protoplasts: further characterization of virus-related protein synthesis. J. gen. Virol. *51:* 373–383 (1980).

Stace-Smith, R.: Comoviruses; in Kurstak, Handbook of plant virus infections and comparative diagnosis, pp. 171–195 (Elsevier/North Holland, Amsterdam 1981).

Stanley, J.; Rottier, P.; Davies, J.W.; Zabel, P.; van Kammen, A.: A protein linked to the 5′ termini of both RNA components of the cowpea mosaic virus genome. Nucl. Acids Res. *5:* 4505–4522 (1978).

Taxonomic status	*English vernacular name*	*International name*
Group	Tobacco ringspot virus group	***NEPOVIRUS*** (185)
Type member	Tobacco ringspot virus (TobRV) (17)	

Main characteristics A. Properties of the Virus Particle

Nucleic acid: Two species of linear positive-sense ssRNA with MW$\simeq 2.8 \times 10^6$ (RNA-1) and $1.3–2.4 \times 10^6$ (RNA-2). The two RNA molecules have little base sequence homology. Each RNA molecule has a poly(A) tract at its 3′ end and a small polypeptide (MW$\simeq 3–6 \times 10^3$) covalently linked, probably to its 5′ end; enzymatic degradation of this polypeptide decreases or abolishes the infectivity of the RNA. The cistron for the genome-linked protein is on RNA-1. 'Satellite' RNA molecules are associated with some strains of some members.
Protein: One coat polypeptide of MW$\simeq 55–60 \times 10^3$; possibly an oligomer or may contain repeated amino acid sequences.
Lipid: None reported.
Carbohydrate: None detected.
Physicochemical properties: Particles usually very stable and, with most members, sediment as three components, T, M and B, respectively containing $\simeq$0, 27–40 and 42–46% RNA by weight with S_{20w} between 49–56, 86–128 and 115–134, and MWs$\simeq$3.2–3.4, 4.6–5.8, and $6.0–6.2 \times 10^6$. Diffusion coefficients of all types of particle are about 1.5×10^{-7} cm^2/s. Buoyant densities in CsCl$\simeq$1.28 (T), 1.43–1.48 (M), and 1.51–1.53 (B) g/cm^3. Satellite RNAs become packaged in helper virus capsids to form additional sedimenting and buoyant density components.
Morphology: All three sedimenting components possess isometric particles$\simeq$28 nm in diameter, often with hexagonal outlines. M particles contain a single molecule of RNA-2, B particles a single molecule of RNA-1; some members have a second type of B particle containing two molecules of RNA-2.
Antigenic properties: Efficient immunogens. Few instances of serological cross-reactivity between members.

B. Replication

Unfractionated RNA induces many local lesions in assay hosts, but separated RNA species induce few or none. RNA-1 can replicate in protoplasts but, in the absence of RNA-2 (which carries the coat protein cistron), no virus particles are produced. RNA-1 presumably carries information for the polymerase function. Virus-induced RNA-dependent RNA polymerase is present in TobRV-infected tissue along with short dsRNA molecules of unknown function. Inhibitor studies indicate that nepovirus proteins are synthesized on cytoplasmic ribosomes. Both RNA species are translated in vitro into large polypeptides approaching in size the theoretical coding capacity; these 'polyproteins' are presumably cleaved in vivo to form the functional proteins.
Characteristic vesiculated inclusion bodies occur in the cytoplasm, usually adjacent to the nucleus. Virus particle antigen accumulates in these structures, which may be the sites of synthesis or assembly of virus components. Newly formed virus particles accumulate in the cytoplasm. They are also commonly found in the plasmodesmata and in single files within tubules in the cytoplasm.

C. Biological Aspects

Host range: Wide. Ringspot symptoms are characteristic, but spotting or mottling symptoms are probably more frequent. Leaves produced later are often symptomless though infected ('recovery'). Symptomless infection is common.

Taxonomic status	*English vernacular name*	*International name*
	Transmission: Readily transmissible experimentally by mechanical inoculation. Seed transmission (via either gamete) is very common. Most members are transmitted by soil-inhabiting longidoroid nematodes, but the vectors of some are unknown. Nematodes retain ability to transmit virus for weeks or months but cease to transmit after moulting. The viruses do not multiply in the vector.	
Other members	*Arabis* mosaic (16) Arracacha A (216) Artichoke Italian latent (176) Artichoke yellow ringspot Blueberry leaf mottle Cherry leaf roll (80) Chicory yellow mottle (132) Cocoa necrosis (173) Crimson clover latent Grapevine Bulgarian latent (186) Grapevine chrome mosaic (103) Grapevine fanleaf (28) *Hibiscus* latent ringspot (233) Lucerne Australian latent (225) Mulberry ringspot (142) Myrobalan latent ringspot (160) Peach rosette mosaic (150) Potato black ringspot (206) Raspberry ringspot (198) Tomato black ring (38) Tomato ringspot (18)	
Possible members	Artichoke vein banding Cherry rasp leaf (159) Satsuma dwarf (208) Strawberry latent ringspot (126) Tomato top necrosis	
Derivation of name	nepo: sigla from *ne*matode, *po*lyhedral to distinguish these viruses from the *Tobravirus* group	

References

Breuning, G.: Plant covirus systems: two component systems; in Fraenkel-Conrat, Wagner, Comprehensive virology, vol. 11, pp. 55–141 (Plenum, New York 1977).

Chu, P.W.S.; Francki, R.I.B.: The chemical subunit of tobacco ringspot virus coat protein. Virology *93:* 398–412 (1979).

Fritsch, C.; Mayo, M.A.; Murant, A.F.: Translation products of genome and satellite RNAs of tomato black ring virus. J. gen. Virol. *46:* 381–389 (1980).

Gallitelli, D.; Rana, G.L.; Di Franco, A.: Il virus della scolorazione perinervale del carciofo (artichoke vein banding virus). Phytopath. Mediterranea *17:* 1–7 (1978).

Kenten, R.H.; Cockbain, A.J.; Woods, R.D.: Crimson clover latent virus – a newly recognised seed-borne virus infecting crimson clover *(Trifolium incarnatum)*. Ann. appl. Biol. *96:* 79–85 (1980).

Mayo, M.A.; Barker, H.; Harrison, B.D.: Polyadenylate in the RNA of five nepoviruses. J. gen. Virol. *43:* 603–610 (1979a).

Mayo, M.A.; Barker, H.; Harrison, B.D.: Evidence for a protein covalently linked to tobacco ringspot virus RNA. J. gen. Virol. *43:* 735–740 (1979b).

Mircetich, S.M.; Sanborn, R.R.; Ramos, D.E.: Natural spread, graft-transmission and possible etiology of walnut blackline disease. Phytopathology *70:* 962–968 (1980).

Murant, A.F.: Nepoviruses; in Kurstak, Handbook of plant virus infections and comparative diagnosis, pp. 197–238 (Elsevier/North Holland, Amsterdam 1981).

Murant, A.F.; Taylor, M.; Duncan, G.H.; Raschké, J.H.: Improved estimate of molecular weight of plant virus RNA by agarose gel electrophoresis and electron microscopy after denaturation with glyoxal. J. gen. Virol. *53:* 321–332 (1981).

Ramsdell, D.C.; Stace-Smith, R.: Physical and chemical properties of blueberry leaf mottle virus. Phytopathology *71:* 468–472 (1981).

Rana, G.L.; Gallitelli, D.; Kyriakopoulon, P.E.; Russo, M.; Martelli, G.P.: Host range and properties of artichoke yellow ringspot virus. Ann. appl. Biol. *96:* 177–185 (1981).

Robinson, D.J.; Barker, H.; Harrison, B.D.; Mayo, M.A.: Replication of RNA-1 of tomato black ring virus independently of RNA-2. J. gen. Virol. *51:* 317–326 (1980).

Taxonomic status	*English vernacular name*	*International name*
Group (monotypic)	Pea enation mosaic virus group	–
Type member	Pea enation mosaic virus (25)	–
Main characteristics	A. Properties of the Virus Particle Nucleic acid: Two molecules of linear positive-sense ssRNA with MWs$\simeq$1.7 and 1.3×10^6. Some strains also contain a third RNA component with MW$\simeq 0.3 \times 10^6$. Protein: Major coat polypeptide, MW$\simeq 22 \times 10^3$; minor polypeptide (MW$= 28 \times 10^3$) associated with aphid transmissibility. Lipid: None. Carbohydrate: None. Physicochemical properties: Particles of two types (B and M) with MWs$\simeq 5.7 \times 10^6$ (B) and$\simeq 4.6 \times 10^6$ (M); density in CsCl$\simeq$1.42 g/cm^3 for B component; M component is disrupted. Density in Cs_2SO_4 = 1.380 g/cm^3 for both components. $S_{20w}\simeq 112$ (B) and $\simeq 99$ (M); particles readily disrupted in neutral chloride salts. Morphology: Polyhedral particles, diameter$\simeq$28 nm. Antigenic properties: Poor immunogen. One or two precipitin lines are formed in gel diffusion tests. B. Replication Replicates in the nucleus. Vesicular cytopathological structures originating from nuclear membranes develop in infected cells. C. Biological Aspects Host range: Narrow host range in plants. Transmission: Transmitted by aphids in a persistent manner. Readily transmissible experimentally by mechanical inoculation, often with loss of aphid transmissibility.	

References

de Zoeten, G.A.; Gaard, G.; Diez, F.B.: Nuclear vesiculation associated with pea enation mosaic virus-infected plant tissue. Virology *48:* 638–647 (1972).

Hull, R.: Particle differences related to aphid-transmissibility of a plant virus. J. gen. Virol. *34:* 183–187 (1977).

Hull, R.: Pea enation mosaic virus; in Kurstak, Handbook of plant virus infections and comparative diagnosis, pp. 239–265 (Elsevier/North Holland, Amsterdam 1981).

Powell, C.A.; de Zoeten, G.A.; Gaard, G.: The localization of pea enation mosaic virus-induced RNA-dependent RNA polymerase in infected peas. Virology *78:* 135–143 (1977).

Taxonomic status	English vernacular name	International name
Family	Nodamura virus group	***NODAVIRIDAE***

Main characteristics

A. Properties of the Virus Particle

Nucleic acid: Two ssRNA molecules, one each of 1.15 and 0.46×10^6 in the same particle; 20.5% of particle by weight. There is no poly(A) 3′ sequence. Both molecules of the isolated RNA are required for infection.

Protein: One major polypeptide of MW 40×10^3, and possibly one of 43×10^3 or two of 39×10^3.

Lipid: Not determined, probably none.

Carbohydrate: Not determined.

Physicochemical properties: MW of virions 8×10^6; $S_{20w} = 135$. Buoyant density in CsCl = 1.34 g/cm^3. Resistant to organic solvents. Stable to pH 3.0. Unstable in presence of chloride ions.

Morphology: Virus particles are roughly spherical, naked nucleocapsids 29 nm in diameter, probably icosahedral.

Antigenic properties: The type species is serologically distinct from other members, some of which are closely inter-related.

B. Replication

The virus replicates in the cytoplasm of cells of several tissues. The two RNA species are translated separately, the larger RNA codes for a protein of MW = 105×10^3, the smaller for a protein of MW 43×10^3. The 43×10^3 polypeptide appears to be a precursor of the major capsid protein.

C. Biological Aspects

Host range: Natural – All species were isolated from insects; from Diptera, Coleoptera and Lepidoptera. Most species are not host-specific. Artificial – Nodamura virus can be grown in various insects and in suckling mice. It replicates in vertebrate and invertebrate cell cultures.

Transmission: Nodamura virus is transmissible to suckling mice by *Aedes aegypti*. Other members will grow in several insect species and invertebrate cells, but not mice or vertebrate cells.

Genus	–	*Nodavirus*
Type species	Nodamura virus	–
Other member	Black beetle virus	
Possible members	Arkansas bee Boolarra Endogenous *Drosophila* line Flock house	
Derivation of name	Nodamura: place in Japan where type species was isolated	

References

Longworth, J. F.: Small isometric viruses of invertebrates. Adv. Virus Res. *23:* 103–157 (1978).
Newman, J. F. E.; Matthews, T.; Omilianowski, D. R.; Saleroni, T.; Kaesberg, P.; Rueckert, R.: *In vitro* translation of the two RNAs of Nodamura virus, a novel mammalian virus with a divided genome. J. Virol. *25:* 78–85 (1978).

Taxonomic status	*English vernacular name*	*International name*
None (possible group)	Velvet tobacco mottle virus group	–
None (possible type member)	Velvet tobacco mottle virus (VTMoV)	–
Main characteristics	A. Properties of the Virus Particle Nucleic acid: Linear ssRNA, MW$\simeq$1.5 $\times$ 10^6, and circular ssRNA, MW$\simeq$1.2 $\times$ 10^5, required for infectivity. Small linear molecules with same base sequence as circular RNA are also encapsidated. Protein: One major polypeptide with MW 30–33 $\times$ 10^3. Lipid: None reported. Carbohydrate: None reported. Physicochemical properties: Density in CsCl$\simeq$1.37 g/cm^3; $S_{20w}$$\simeq$115. Morphology: Polyhedral particles; $\simeq$30 nm in diameter. Antigenic properties: Very efficient immunogens. Single precipitin line in gel diffusion tests. Serological relationships between some members. B. Replication Particles located in nuclei, cytoplasm and vacuoles; vesicles of various sizes containing electron-dense strands in cytoplasm. C. Biological Aspects Host range: Each virus has narrow host range. Transmission: Transmitted by coccinellid beetles and myrids. Readily transmissible by mechanical inoculation.	
Other members	Lucerne transient streak *Solanum nodiflorum* mottle Subterranean clover mottle	

References

Gould, A.R.; Francki, R.I.B.; Randles, J.W.: Studies on encapsidated viroid-like RNA. IV. Requirement for infectivity and specificity of two RNA components from velvet tobacco mottle virus. Virology *110:* 420–426 (1981).

Gould, A.R.; Hatta, T.: Studies on encapsidated viroid-like RNA. III. Comparative studies on RNAs isolated from velvet tobacco mottle virus and solanum nodiflorum mottle virus. Virology *109:* 137–147 (1981).

Greber, R.S.: Some characteristics of *Solanum nodiflorum* mottle virus – a beetle-transmitted isometric virus from Australia. Aust. J. biol. Sci. *34:* 369–378 (1981).

Randles, J.W.; Davies, C.; Hatta, T.; Gould, A.R.; Francki, R.I.B.: Studies on encapsidated viroid-like RNA. I. Characterization of velvet tobacco mottle virus. Virology *108:* 111–122 (1981).

Tien, P.; Davies, C.; Hatta, T.; Francki, R.I.B.: Viroid-like RNA encapsidated in lucerne transient streak virus. FEBS Lett. (in press).

Taxonomic status	*English vernacular name*	*International name*
Group	Tobacco rattle virus group	***TOBRAVIRUS***
Type member	Tobacco rattle virus (PRN isolate) (12)	–
Main characteristics	A. Properties of the Virus Particle Nucleic acid: Two strands of linear positive-sense ssRNA with MWs of 2.4×10^6 (RNA-1) and $0.6–1.4 \times 10^6$ (RNA-2), the size of the latter depending on the isolate; 5′ terminus of RNA-2 has the sequence $m^7G^{5\prime}ppp^{5\prime}Ap$... RNA-1 does not possess this termination. RNA-1 is infective; RNA-2 is not infective but it contains cistron for capsid protein; both RNAs are required for production of progeny long (L) and shor pt (S)articles. Protein: One coat polypeptide; $MW \simeq 22 \times 10^3$. Lipid: None. Carbohydrate: None. Physicochemical properties: $MWs = 48–50 \times 10^6$ (L) and $11–29 \times 10^6$ (S); $S_{20w} = 296–306$ (L) and 155–245 (S); density in $CsCl = 1.306–1.324$ g/cm^3. Particles quite stable. Morphology: Tubular particles with helical symmetry with pitch of 2.5 nm; diameter = 21.3–23.1 nm (electron microscopy) or 20.5–22.5 nm (X-ray). RNA-1 and RNA-2 contained in tubular particles of 180–215 nm length (L) and 46–114 nm length (S), the latter depending on the isolate. Antigenic properties: One moderately immunogenic; considerable antigenic heterogeneity between isolates. B. Replication Accumulation of virus particles sensitive to cycloheximide but not to chloramphenicol, suggesting cytoplasmic ribosomes are involved in viral protein synthesis; L particles accumulate in early part of infection cycle, while S particles tend to accumulate in the later stages; isolates unable to produce nucleoprotein particles (NM isolates) are obtained from inocula containing only L particles; such isolates are also found in naturally infected plants. C. Biological Aspects Host range: Wide, including 50 monocotyledonous and dicotyledonous families. Transmission: Primarily by nematodes (*Paratrichodorus* and *Trichodorus* spp.) in which the virus persists, but there is no evidence of replication; by seed; and by contact, but with difficulty in some instances.	
Other member	Pea early-browning virus (120)	
Possible member	Peanut clamp virus	
Derivation of name	tobra: sigla from *to*bacco *ra*ttle	

References

Harrison, B.D.; Robinson, D.J.: The tobraviruses. Adv. Virus Res. *23:* 25–77 (1978).

Harrison, B.D.; Robinson, D.J.: Tobraviruses; in Kurstak, Handbook of plant virus infections and comparative diagnosis, pp. 515–540 (Elsevier/North Holland, Amsterdam 1981).

Taxonomic status	*English vernacular name*	*International name*
Group	Cucumber mosaic virus group	***CUCUMOVIRUS***
Type member	Cucumber mosaic virus (CMV) (S isolate) (1, 213)	

Main characteristics

A. Properties of the Virus Particle

Nucleic acid: Three molecules of linear positive-sense ssRNA; MWs$\simeq$1.27 (RNA-1), 1.13 (RNA-2) and 0.82×10^6 (RNA-3); 0.35×10^6 MW coat protein mRNA (RNA-4) is also encapsidated. 5′ termini of the four RNAs have the sequence: $m^7G^{5'}ppp^{5'}Np$... The 3′ termini have a tRNA-like structure which accepts tyrosine.

Protein: Single coat polypeptide, MW$\simeq 24 \times 10^3$.

Lipid: None reported.

Carbohydrate: None reported.

Physicochemical properties: MW$\simeq 6 \times 10^6$; $S_{20w}\simeq 99$; density in CsCl$\simeq$1.37 g/cm^3; particles readily disrupted in neutral chloride salts and by SDS; particles sensitive to RNase.

Morphology: Polyhedral particles, $\simeq$29 nm in diameter, with T=3 surface lattice symmetry. Although all particles have approximately the same S_{20w} ($\simeq$99), three particles exist, one containing one molecule of RNA-1, one containing one molecule of RNA-2 and one containing one molecule each of RNA-3 and RNA-4 (some isolates of CMV and peanut stunt virus also encapsidate short linear ss-satellite RNAs, MW$\simeq 1 \times 10^5$).

Antigenic properties: Poor immunogens. Serological reactions complicated by sensitivity of virus particles to salts. Distant serological relationships among members.

B. Replication

The RNAs of CMV can each be translated in vitro to yield 4 major proteins; RNAs 1, 2, 3 and 4 code for proteins of MW$\simeq$105, 120, and 34×10^3 and coat protein, respectively. RNAs replicate via corresponding negative-sense strands. Virus particles assemble in the cytoplasm and accumulate there as scattered particles. Sometimes, virus particles also occur in nuclei and vacuoles, rarely forming crystals. Small vesicles associated with the tonoplast may be the sites of RNA replication.

C. Biological Aspects

Host range: Narrow.

Transmission: Readily transmissible experimentally by mechanical inoculation. Seed transmission in several host plants. Transmitted by aphids in non-persistent manner.

Other members	Peanut stunt (92) Tomato aspermy (79)	
Probable member	Cowpea ringspot	
Derivation of name	cucumo: sigla from *cucum*ber *mo*saic	

References

Hatta, T.; Francki, R.I.B.: Cytopathic structures associated with tonoplast of plant cells infected with cucumber mosaic and tomato aspermy viruses. J. gen. Virol. *53:* 343–346 (1981).

Kaper, J. M.; Waterworth, H. E.: Cucumoviruses; in Kurstak, Handbook of plant virus infections and comparative diagnosis, pp. 257–332 (Elsevier/North Holland, Amsterdam 1981).
Phatak, H. C.; Diaz-Ruiz, J. R.; Hull, R.: Cowpea ringspot virus: a seed transmitted cucumovirus. Phytopath. Z. *87:* 132–142 (1976).

Taxonomic status	*English vernacular name*	*International name*
Group	Brome mosaic virus group	***BROMOVIRUS*** (215)
Type member	Brome mosaic virus (BMV) (3, 180)	–
Main characteristics	A. Properties of the Virus Particle Nucleic acid: Three molecules of linear positive-sense ssRNA with MWs$\simeq$1.1 (RNA-1), 1.0 (RNA-2), and 0.7×10^6 (RNA-3); 0.3×10^6 MW coat protein mRNA (RNA-4) is also encapsidated. 5′ termini of the RNAs have the sequence: $m^7G^{5'}ppp^{5'}Gp$... The 3′termini have a tRNA-like structure which accepts tyrosine. Protein: Single coat polypeptide, MW$\simeq 20 \times 10^3$. Lipid: None. Carbohydrate: None. Physicochemical properties: MW$\simeq 4.6 \times 10^6$; $S_{20w}\simeq 85$; buoyant density in CsCl$\simeq$1.35 g/cm^3; alkaline pH (7–8) induces swelling of virus particles, and their susceptibility to proteases and ribonuclease can be prevented by Ca^{++}; particles readily disrupted in chloride salts and by sodium dodecyl sulfate. Morphology: Polyhedral particles, $\simeq$26 nm in diameter, with icosahedral T=3 surface lattice symmetry. Although all particles have approximately the same S_{20w} ($\simeq$85), three different particles exist, one containing one molecule of RNA-1, one containing one molecule of RNA-2 and one containing one molecule each of RNA-3 and RNA-4. Antigenic properties: Relatively poor immunogens. Serological reactions complicated by instability. Distant serological relationships among members. B. Replication The RNAs of BMV and cowpea chlorotic mottle virus can each be translated in vitro to yield 4 major proteins; RNAs 1, 2, 3 and 4 code for proteins of MW$\simeq$112, 107, 33 and 20×10^3, respectively. Each RNA replicates via corresponding negative-sense strand. Virus assembles in cytoplasm. Granular inclusions appear in cytoplasm. Crystalline arrays of virus particles are sometimes seen. Particles can be seen in both cytoplasm and nuclei of old infected cells. C. Biological Aspects Host range: Narrow. Transmission: Readily transmissible experimentally by mechanical inoculation. Some members transmitted by beetles.	
Other members	Broad bean mottle (101) Cowpea chlorotic mottle (49)	
Possible member	*Melandrium* yellow fleck (236)	
Derivation of name	bromo: sigla from *bro*me *mo*saic; also, from plant generic name *Bromus*, brome	

References

Bancroft, J.B.; Horne, R.W.: Bromovirus (brome mosaic virus) group; in Maramorosch, The atlas of insect and plant viruses, pp. 287–302 (Academic Press, New York 1977).

Davies, J.W.; Verduin, B.J.M.: *In vitro* synthesis of cowpea chlorotic mottle virus polypeptides. J. gen. Virol. *44:* 545–549 (1979).

Lane, L.C.: Bromoviruses; in Kurstak, Handbook of plant virus infections and comparative diagnosis, pp. 333–376 (Elsevier/North Holland, Amsterdam 1981).

Loesch-Fries, L.S.; Hall, T.C.: Synthesis, accumulation and encapsidation of individual brome mosaic virus RNA components in barley protoplasts. J. gen. Virol. *47:* 323–332 (1980).

Pfeiffer, P.: Changes in the organization of bromegrass mosaic virus in response to cation binding as probed by changes in susceptibility to degradative enzymes. Virology *102:* 54–61 (1980).

Rybicke, E.P.; von Wechman, M.B.: The serology of the bromoviruses. I. Serological interrelationships of the bromoviruses. Virology *109:* 391–402 (1981).

Taxonomic status	*English vernacular name*	*International name*
Group	Tobacco streak virus group	***ILARVIRUS***
Type member	Tobacco streak virus (TSV) (44)	–
Main characteristics	A. Properties of the Virus Particle Nucleic acid: Three molecules of linear positive-sense ssRNA; MW$\simeq$1.1 (RNA-1), 0.9 (RNA-2) and 0.7×10^6 (RNA-3); 0.3×10^6 MW coat protein mRNA (RNA-4) is also encapsidated. Protein: Single coat polypeptide, MW$\simeq 25\times10^3$. Lipid: None reported. Carbohydrate: None reported. Physicochemical properties: Several particle types, S_{20w} ranging from 80 to 120; density of all particle types$\simeq$1.36 g/cm^3 in CsCl; particles readily disrupted in neutral chloride salts and by SDS. Morphology: Particles are quasi-isometric or occasionally bacilliform. Particles of different components, although differing in size, are mostly 26–35 nm in diameter. Antigenic properties: Weakly to moderately immunogenic. Serological relationship among some members. B. Replication Besides RNAs 1–3, coat protein or RNA-4 is required for infectivity. Coat protein of most ilarviruses (and also of alfalfa mosaic virus) are interchangeable in this respect. For some members it has been shown that RNAs 1 and 2 can be translated in vitro into proteins of a MW corresponding to the total genetic information present in these RNAs. RNA-3 directs the synthesis of protein, MW$\simeq 35\times10^3$, and RNA-4 directs the synthesis of coat protein. C. Biological Aspects Host range: Wide Transmission: Readily transmissible experimentally by mechanical inoculation. Some viruses transmitted by seeds and by pollen to flower-bearing plants.	
Other members	Apple mosaic (=Danish plum line pattern, hop A and rose mosaic) (83) *Citrus* leaf rugose (164) *Citrus* variegation Elm mottle (139) Lilac ring mottle (201) North American plum line pattern Prune dwarf (19) *Prunus* necrotic ringspot (=cherry rugose mosaic and hop B) (5) Spinach latent Tulare apple mosaic (42) (*Note:* Black raspberry latent virus appears to be a strain of tobacco streak virus, the type member of the group.)	
Derivation of name	ilar: sigla from *i*sometric *la*bile *r*ingspot	

References

Bos, L.; Huttinga, H.; Maat, D.Z.: Spinach latent virus, a new ilarvirus seed-borne in *Spinacia oleracea*. Neth. J. Plant Pathol. *86:* 79–98 (1980).

Fulton, R.W.: Ilarviruses; in Kurstak, Handbook of plant virus infections and comparative diagnosis, pp. 257–332 (Elsevier/North Holland, Amsterdam 1981).

Huttinga, H.; Mosch, W.H.M.: Lilac ring mottle virus: a coat protein-dependent virus with a tripartite genome. Acta horticult. *59:* 113–118 (1976).

van Vloten-Doting, L.: Coat protein is required for infectivity of tobacco streak virus: biological equivalence of the coat proteins of tobacco streak and alfalfa mosaic viruses. Virology *65:* 215–225 (1975).

Taxonomic status	*English vernacular name*	*International name*
Group (monotypic)	Alfalfa mosaic virus group	–
Type member	Alfalfa mosaic virus (AMV) (46, 229)	–
Main characteristics	A. Properties of the Virus Particle	

Nucleic acid: Three molecules of linear positive-sense ssRNA with MWs$\simeq$1.1 (RNA-1), 0.8 (RNA-2) and 0.7×10^6 (RNA-3); 0.3×10^6 MW coat protein mRNA (RNA-4) is also encapsidated. 5′ termini of the four RNAs have the sequence $m^7G^{5'}ppp^{5'}Gp$...
Protein: One coat polypeptide, MW$\simeq 24 \times 10^3$.
Lipid: None reported.
Carbohydrate: None reported.
Physicochemical properties: Particles of at least four sizes (B, M, Tb and Ta) with MWs ranging from 7.3 to 3.7×10^6; $S_{20w} \simeq 99$ (B), 88 (M), 75 (Tb) and 68 (Ta); density in Cs_2SO_4 $\simeq$1.28 g/cm^3 (components differ slightly in banding densities); particles disrupted in neutral chloride salts and by sodium dodecyl sulfate; particles sensitive to ribonuclease at pH 6–7, but there is no apparent swelling of particles.
Morphology: Bacilliform particles 58 × 18 nm (B), 48 × 18 nm (M), 36 × 18 nm (Tb) and an ellipsoidal one, approximately 28 × 18 nm (Ta). The three largest particles contain a single RNA molecule each: RNA-1 (B), RNA-2 (M), RNA-3 (Tb); Ta contains two molecules of RNA-4.
Antigenic properties: Poor immunogens.

B. Replication
Besides RNAs 1–3, coat protein or RNA-4 is required for infectivity. Coat proteins from most ilarviruses are also able to activate the AMV genome. The two largest RNAs can be translated in vitro into proteins of MW corresponding to total genetic information present in these RNAs. RNA-3 directs mainly the synthesis of coat protein. Virus particles accumulate in the cytoplasm and sometimes in vacuoles, either scattered or as whorled aggregates.

C. Biological Aspects
Host range: Relatively wide.
Transmission: Readily transmissible experimentally by mechanical inoculation. Seed transmission in some plants. Transmitted by aphids in nonpersistent manner.

References

Lane, L.C.: The nucleic acids of multipartite, defective, and satellite plant viruses; in Hall, Davies, Nucleic acids in plants, vol. 2, pp. 65–110 (CRC Press, West Palm Beach 1979).
van Vloten-Doting, L.; Francki, R.I.B.; Fulton, R.W.; Kaper, J.M.; Lane, L.C.: Tricornaviridae – a proposed family of plant viruses with tripartite, single-stranded RNA genomes. Intervirology *15:* 198–203 (1981).

Taxonomic status	*English vernacular name*	*International name*
Group	Barley stripe mosaic virus group	***HORDEIVIRUS***
Type member	Barley stripe mosaic virus (BSMV) (68)	–
Main characteristics	A. Properties of the Virus Particle Nucleic acid: 2–4 molecules of linear positive-sense ssRNA; number of molecules depending on strain; 2 or 3 RNA components necessary for infection. MWs range from 1 to 1.5 $\times 10^6$. RNA is polyadenylated at 3′ end with short (15–20) nucleotide tracts of A; the 5′-termini of two of the three RNAs (Norwich strain) have the sequence $m^7G^{5\prime}ppp^{5\prime}Gp$... Protein: Single polypeptide, MW$\simeq 21 \times 10^3$. Lipid: None. Carbohydrate: Protein is glycosylated. Physicochemical properties: One component has MW$\simeq 26 \times 10^6$; $S_{20w} = 185$. Additional components with $S_{20w} \simeq 178$ and 200 also present in purified preparations. Morphology: Elongated rigid particles about 20 nm in diameter and 100–150 nm long, helically symmetrical with pitch$\simeq$2.5 nm. Antigenic properties: Efficient immunogens. B. Replication RF RNAs corresponding to all viral ssRNAs can be isolated from infected plants. Virus particles accumulate in both cytoplasm and nuclei, most being in the cytoplasm. C. Biological Aspects Host range: Narrow host range, mostly among Gramineae. Transmission: By mechanical inoculation and through seed.	
Other members	*Lychnis* ringspot *Poa* semilatent	
Derivation of name	hordei: from Latin *hordeum*, 'barley'	

References

Agranovsky, A.A.; Dolja, V.V.; Davsan, V.M.; Atabekov, J.G.: Detection of polyadenylate sequences in RNA components of barley stripe mosaic virus. Virology *91:* 95–105 (1978).

Agranovsky, A.A.; Dolja, V.V.; Kagramanova, V.K.; Atabekov, J.G.: The presence of a cap structure at the 5′-end of barley stripe mosaic virus RNA. Virology *95:* 208–210 (1979).

Gumpf, D.J.; Cunningham, D.S.; Heick, J.A.; Shannon, L.M.: Amino acid sequence in the proteolytic glycopeptide of barley stripe mosaic virus. Virology *78:* 328–330 (1977).

Jackson, A.O.; Lane, L.C.: Hordeiviruses; in Kurstak, Handbook of plant virus infections and comparative diagnosis, pp. 565–625 (Elsevier/North Holland, Amsterdam 1981).

Lane, L.C.: The components of barley stripe mosaic and related viruses. Virology *58:* 323–333 (1974).

Palomar, M.K.; Brakke, M.K.; Jackson, A.O.: Base sequence homology in the RNAs of barley stripe mosaic virus. Virology *77:* 471–480 (1977).

Some Unclassified Viruses and Virus-Like Agents

For many viruses and virus-like agents, the data on characterization are inadequate at present to allow for sound classification. Some of the most important and interesting of these are listed below:

1. Human hepatitis B virus and other similar viruses
2. Marburg/Ebola
3. Agents of scrapie (and transmissible mink encephalopathy), Kuru, and Creutzfeldt-Jakob diseases
4. Viroids
5. Borna disease virus
6. Agent of roseola infantum (or exanthem subitum)
7. Agent of cat-scratch disease
8. Satellite viruses in plants
9. Satellite RNAs in plants and fungi

Index of Virus Names[1]

[1] Approved names for virus families, subfamilies and plant virus groups are given in bold-face italic type. Approved names for virus genera, subgenera, and species are given in italic type. 'Probable' and 'possible' members. Host generic and specific names are given in italic type.